HANDBOOK OF
ENERGY AUDITS

Fifth Edition

HANDBOOK OF ENERGY AUDITS

Fifth Edition

Albert Thumann, P.E., C.E.M.

Published by
THE FAIRMONT PRESS, INC.
700 Indian Trail
Lilburn, GA 30047

Library of Congress Cataloging-in-Publication Data

Thumann, Albert.
Handbook of energy audits / Albert Thumann.--5th ed.
   p. cm.
   Includes index.
   ISBN 0-88173-294-X
   1. Energy auditing--Handbooks, manuals, etc. I. Title.
   TJ163.245.T48   1998   658.2'6--dc21   98-11968
                                               CIP

Published by The Fairmont Press, Inc.
700 Indian Trail
Lilburn, GA 30247

Printed in the United States of America

10 9 8 7 6 5 4 3 2 1

ISBN 0-88173-294-X FP

ISBN 0-13-975202-1 PH

While every effort is made to provide dependable information, the publisher, authors, and
editors cannot be held responsible for any errors or omissions.

Distributed by Prentice Hall PTR
Prentice-Hall, Inc.
A Simon & Schuster Company
Upper Saddle River, NJ 07458

Prentice-Hall International (UK) Limited, London
Prentice-Hall of Australia Pty. Limited, Sydney
Prentice-Hall Canada Inc., Toronto
Prentice-Hall Hispanoamericana, S.A., Mexico
Prentice-Hall of India Private Limited, New Delhi
Prentice-Hall of Japan, Inc., Tokyo
Simon & Schuster Asia Pte. Ltd., Singapore
Editora Prentice-Hall do Brasil, Ltda., Rio de Janeiro

*This book is dedicated to one of the
first energy conservationists and his wife,
my parents.*

Preface

Handbook of Energy Audits, now published in its fifth edition, has become a standard and essential reference for engineers and decision makers throughout business, industry, institutions and government. It has been used by more than 9,000 professionals to date.

The role of the energy audit has changed a great deal since the publication of the first edition of *Handbook of Energy Audits* in 1979. The National Energy Conservation Policy Act of 1978 led to the first wide-scale use of energy audits by mandating they be performed as a part of technical assistance programs for schools and hospitals. In order to receive funds available through matching-grants programs, an energy audit was required to identify potential energy conservation and savings opportunities prior to project implementation.

As we enter the new millennium, the role of energy audits has greatly expanded. Today many large energy consumers are contracting with energy services providers to implement energy projects. This arrangement necessitates the establishing of baseline energy usage, as well as the quantification of savings resulting from project implementation. Accurate and complete energy audits are essential as a means to assess and verify a project's success at meeting contracted goals.

The auditing process has also entered a new domain as restructuring of the energy marketplace takes hold, and energy consumers have new choices. The establishing of specific energy use parameters determined through auditing of utility bills can provide a valuable instrument for negotiating more competitive rates. A new chapter provides guidance to help you set effective energy procurement strategies as we move forward into more far-reaching deregulation of both electricity and natural gas.

This book remains a basic primer to get all professionals up to speed on the most effective tool to assure maintaining the competitive edge...energy auditing.

Albert Thumann, P.E., C.E.M.

Contents

1

What Is An Energy Audit?

Energy audits mean different things to different individuals. Lacking a clear definition, the term "energy audit" has in itself caused confusion. This chapter reviews the background of energy audits and defines how they are presently being used.

WHERE IT STARTED

The term energy audits was used in the *Federal Register,* Vol. 42, No. 25, June 29, 1977, dealing with Energy Audit Procedures. For each State Energy Office to qualify for financial assistance, they had to submit a Supplemental State Energy Conservation Program (SSEP) containing procedures for:

(a) Carrying out a continuing public education effort on implementing energy conservation measures.
(b) Insuring effective inter-governmental conditions.
(c) Encouraging and carrying out energy audits for building and industrial plants.

In addition to energy audits enabling the state to qualify for financial funding, energy audits were also to be used for "loan guarantee" or verification audits. Thus the audit terminology developed with financial assistance and funding in mind.

By abstracting several statements from the first generation procedures, energy audits may be defined as follows:

"The energy audit serves to identify all of the energy streams into a facility and to quantify energy use according to discrete functions.

"An energy audit may be considered as similar to the monthly closing statement of an accounting system. One series of entries consists of amounts of energy which were consumed during the month in the form of electricity, gas, fuel, oil, steam, and the second series lists how the energy was used; how much for lighting, in air-conditioning, in heating, in process, etc. The energy audit process must be carried out accurately enough to identify and qualify the energy and cost savings that are likely to be realized through investment in an energy savings measure."

TYPES OF ENERGY AUDITS

The simplest definition for an energy audit is as follows: An energy audit serves the purpose of identifying where a building or plant facility uses energy and identifies energy conservation opportunities.

There is a direct relationship to the cost of the audit (amount of data collected and analyzed) and the number of energy conservation opportunities to be found. Thus, a first distinction is made between cost of the audit which determines the type of audit to be performed.

The second distinction is made between the type of facility. For example, a building audit may emphasize the building envelope, lighting, heating, and ventilation requirements. On the other hand, an audit of an industrial plant emphasizes the process requirements.

Most energy audits fall into three categories or types, namely, Walk-Through, Mini-Audit, or Maxi-Audit.

Walk-Through—This type of audit is the least costly and identifies preliminary energy savings. A visual inspection of the facility is made to determine maintenance and operation energy saving opportunities plus collection of information to determine the need for a more detailed analysis.

Mini-Audit—This type of audit requires tests and measurements to quantify energy uses and losses and determine the economics for changes.

Maxi-Audit—This type of audit goes one step further than the mini-audit. It contains an evaluation of how much energy is used for each function such as lighting, process, etc. It also requires a model analysis, such as a computer simulation, to determine energy use patterns and predictions on a year-round basis, taking into account such variables as weather data.

MATCHING GRANTS PROGRAM FOR SCHOOLS AND HOSPITALS

BACKGROUND

The Matching Grants Program for Schools and Hospitals is a voluntary program to assist these non-profit institutions in saving energy.

Authorized by Part 1, Title III of the National Energy Conservation Policy Act of 1978 (NECPA) and administered by the Department of Energy's Institutional Conservation Programs (ICP) through state energy offices, the ICP provides grants for detailed energy analyses and for installation of energy-saving capital improvements to eligible institutions. Participation in the grant program requires a 50 percent match of funds from recipient institutions, except in hardship cases.

PROGRAM ACTIVITIES

NECPA divided the program into two main phases: Phase I, Preliminary Energy Audits and Energy Audits, and Phase II, Technical Assistance and Energy Conservation Measures. Funding for Phase I activities has expired; however, the program still administers the Technical Assistance/Energy Conservation Measure phase.

The Preliminary Energy Audit (PEA) was a data-gathering activity to determine basic information about the number, size, type, and rate of energy use in eligible buildings and the actions taken to conserve energy. The Energy Audit (EA) consisted of an on-site visit which confirmed and expanded upon information gathered in the PEA and provided further information to establish a "building profile."

The Technical Assistance (TA) Audit consists of detailed engineering analyses and reports specific costs, payback periods, and projected energy savings possible from the purchase and installation of devices or systems, or from modification to building operations and the building shell. Examples of this type of activity would be the installation of storm windows, insulation, solar hot water or space heating systems, or automatic setback devices.

Energy Conservation Measure (ECM) grants provide for the actual purchase and installation of energy measures recommended as a result of the TA Audit.

Details of the Technical Assistance Audit are illustrated in Figure 1-1.

CONTENTS OF PROGRAM

(a)   A technical assistance program shall be conducted by a qualified technical assistance analyst, who shall consider all possible energy conservation measures for a building, including solar or other renewable resource measures. A technical assistance program shall include a detailed engineering analysis to identify the estimated costs of, and the energy and cost savings likely to be realized from, implementing each identified energy conservation maintenance and operating procedure. A technical assistance program shall also identify the estimated cost of, and the energy and cost savings likely to be realized from, acquiring and installing each energy conservation measure, including solar and other renewable resource measures, that indicate a significant potential for saving energy based upon the technical assistance analyst's initial consideration.

(b)   At the conclusion of a technical assistance program, the technical assistance analyst shall prepare a final report which shall include—

(1)   A description of building characteristics and energy data including—

(i)   The results of the preliminary energy audit and energy audit (or its equivalent) of the building;

(ii)   The operating characteristics of energy using systems; and

(iii)   The estimated remaining useful life of the building;

(2)   An analysis of the estimated energy consumption of the building, by fuel type (in total Btu's and Btu/sq. ft./yr.), at optimum efficiency (assuming implementation of all energy conservation maintenance and operating procedures);

(3)   An evaluation of the building's potential for solar conversion.

(4)   A listing of any known local zoning ordinances and building codes which may restrict the installation of solar systems;

Figure 1-1. Detailed Contents of a Technical Assistance Energy Audit

(5) A description and analysis of all recommendations, if any, for acquisition and installation of energy conservation measures, including solar and other renewable resource measures, setting forth—

(i) A description of each recommended energy conservation measure;

(ii) An estimate of the cost of design, acquisition and installation of each energy conservation measure;

(iii) An estimate of the useful life of each energy conservation measure;

(iv) An estimate of increases or decreases in maintenance and operating costs that would result from each energy conservation measure, if any:

(v) An estimate of the salvage value or disposal cost of each energy conservation measure at the end of its useful life, if any;

(vi) An estimate of the annual energy and energy cost savings (using current energy prices) expected from the acquisition and installation of each energy conservation measure. In calculating the potential energy cost savings of each recommended energy conservation measure, including solar or other renewable resource measure, technical assistance analysts shall—

(A) Assume that all energy savings obtained from energy conservation maintenance and operating procedures have been realized;

(B) Calculate the total energy and energy cost savings, by fuel type, expected to result from the acquisition and installation of all recommended energy conservation measures, taking into account the interaction among the various measures; and

(C) Calculate that portion of the total energy and energy cost savings, as determined in (B) above, attributable to each individual energy conservation measure.

(vii) The simple payback period of each recommended energy conservation measure, taking into account the interactions among the various measures. The simple payback period is calculated by dividing the estimated total cost of the measure, (5)(ii), by the estimated annual cost saving accruing from the measure. For the purposes of ranking applications, the simple payback period shall be calculated using the cost savings resulting from energy savings only, determined on the basis of current energy prices. The estimated cost of the measure shall be the total cost for design and other professional services (excluding costs of a technical assistance program), if any, and acquisition and installation costs. Other economic analyses, such as life-cycle costing, which consider all costs and cost savings, such as maintenance costs and/or savings, resulting from an energy conservation measure, are recommended, but not required, for use by the institution in its decision-making process;

(6) A listing of energy use and cost data for each fuel type used for the prior 12-month period.

(7) A signed and dated certification that the technical assistance program has been conducted in accordance with the requirements of this section and the grant application and that the data presented is accurate to the best of the technical assistance analyst's knowledge.

Figure 1-1. Concluded

THE CERTIFIED ENERGY
MANAGERS PROGRAM (CEM)

In order to help identify qualified professionals who perform energy audits and are responsible for energy management, the Association of Energy Engineers (AEE) created in 1982 the CEM program. More than 1100 individuals have been certified.

WHAT IT TAKES TO APPLY

The prerequisites needed to qualify for certification reflect a flexible attitude toward the ratio of education to practical experience. However, candidates must meet one of the following sets of criteria:

A minimum of three full years of experience in energy engineering or energy management for those who are engineering graduates or Registered Professional Engineers.

OR

A minimum of five to eight years in energy engineering or energy management for graduates with business or related degrees or 2-year technical degree.

OR

A minimum of ten full years in energy engineering or energy management.

ABOUT THE CEM EXAMINATION

Applicants must take a 4-hour, multiple-choice, open-book exam. Candidates may select sections based on personal expertise—for example, maintenance, management, energy analysis, energy management systems. Sample questions and a self-study guide are available from AEE. Actual test questions are framed to ascertain both specific knowledge and practical expertise.

2

Energy Accounting And Analysis

ENERGY USE PROFILES

The energy audit process for a building emphasizes building envelope, heating and ventilation, air-conditioning, plus lighting functions. For an industrial facility the energy audit approach includes process consideration. Figures 2-1 through 2-3 illustrate how energy is used for a typical industrial plant. It is important to account for total consumption, cost, and how energy is used for each commodity such as steam, water, air and natural gas. This procedure is required to develop the appropriate energy conservation strategy.

The top portion of Figure 2-1 illustrates how much energy is used by fuel type and its relative percentage. The pie chart below shows how much is spent for each fuel type. Using a pie-chart representation or nodal flow diagram can be very helpful in visualizing how energy is being used.

Figure 2-2 on the other hand shows how much of the energy is used for each function such as lighting, process, and building heating and ventilation. Pie charts similar to the right-hand side of the figure should be made for each category such as air, steam, electricity, water and natural gas.

Figure 2-3 illustrates an alternate representation for the steam distribution profile.

Several audits are required to construct the energy use profiles, such as:

Envelope Audit—This audit surveys the building envelope for losses or gains due to leaks, building construction, doors, glass, lack of insulation, etc.

7

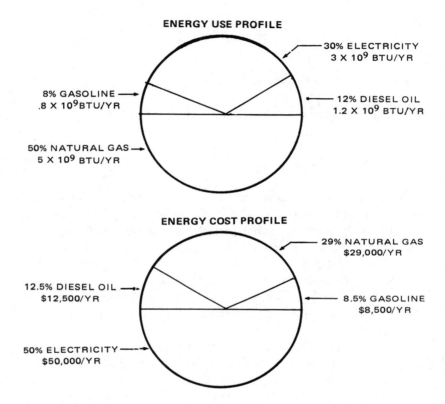

Figure 2-1. Energy Use and Cost Profile

Functional Audit—This audit determines the amount of energy required for a particular function and identifies energy conservation opportunities. Functional audits include:

- Heating, ventilation and air-conditioning
- Building
- Lighting
- Domestic hot water
- Air distribution

Process Audit—This audit determines the amount of energy required for each process function and identifies energy conservation opportunities. Process functional audits include:

- Process machinery

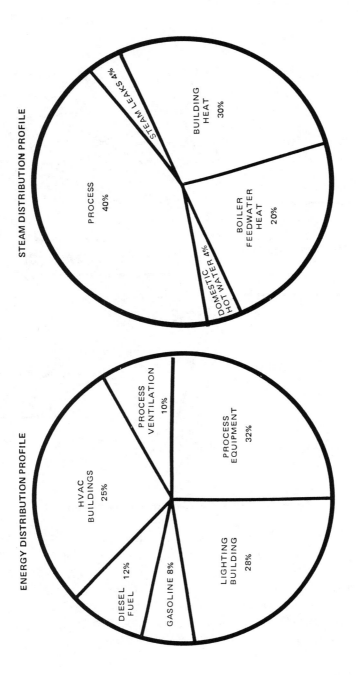

Figure 2-2. Energy Profile by Function

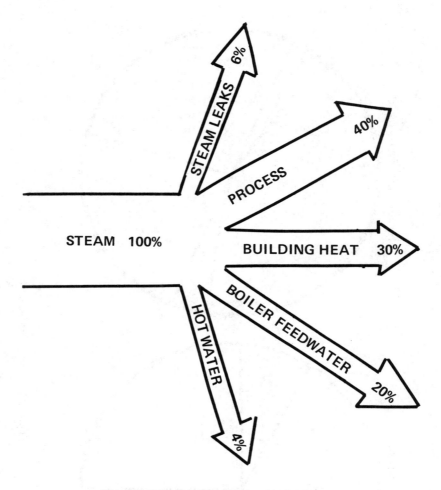

Figure 2-3. Steam Distribution Nodal Diagram

- Heating, ventilation and air-conditioning process
- Heat treatment
- Furnaces

Transportation Audit—This audit determines the amount of energy required for forklift trucks, cars, vehicles, trucks, etc.

Utility Audit—This audit analyzes the monthly, daily or yearly energy usage for each utility.

ENERGY USERS

Energy use profiles for several end-users are summarized in Tables 2-1 through 2-11.

Table 2-1. Energy Use in Apartment Buildings

	Range (%)	Norms (%)
Environmental Control	50 to 80	70
Lighting and Wall Receptacles	10 to 20	15
Hot Water	2 to 5	3
Special Functions		
Laundry, Swimming Pool, Restaurants, Parking, Elevators, Security Lighting	5 to 20	10

Table 2-2. Energy Use in Bakeries

Housekeeping Energy	Percent
Space Heating	21.5
Air Conditioning	1.6
Lighting	1.4
Domestic Hot Water	1.8
TOTAL	26.3
Process Energy	Percent
Baking Ovens	49.0
Pan Washing	10.6
Mixers	4.1
Freezers	3.3
Cooking	2.0
Fryers	1.8
Proof Boxes	1.8
Other Processes	1.1
TOTAL	73.7

Data for a 27,000 square foot bakery in Washington, D.C.

Table 2-3. Energy Use in Die Casting Plants

Housekeeping Energy	Percent
Space Heating	24
Air Conditioning	2
Lighting	2
Domestic Hot Water	2
TOTAL	30
Process Energy	Percent
Melting Hearth	30
Quiet Pool	20
Molding Machines	10
Air Compressors	5
Other Processes	5
TOTAL	70

Table 2-4. Energy Use in Hospital Buildings

	Range (%)	Norms (%)
Environmental Control	40 to 65	58
Lighting and Wall Receptacles	10 to 20	15
Laundry	8 to 15	12
Food Service, Kitchen Operations	5 to 10	7
Medical Equipment, Sterilization, Incinerator, Parking, Elevators, Security Lighting	5 to 15	8

Table 2-5. Energy Use in Hotels and Motels

	Range (%)	Norms (%)
Space Heating	45 to 70	60
Lighting	5 to 15	11
Air Conditioning	3 to 15	10
Refrigeration	0 to 10	4
Special Functions	5 to 20	15
Laundry, Kitchen, Restaurant, Swimming Pool, Garage, Security Lighting, Hot Water		

Table 2-6. Energy Use in Retail Stores

	Range (%)	Norms (%)
HVAC	20 to 50	30
Lighting	40 to 75	60
Special Functions	5 to 20	10
Elevators, General Power, Parking Security Lighting, Hot Water		

Table 2-7. Energy Use in Restaurants

	Table Restaurant Norms (%)	Fast Food Restaurant Norms (%)
HVAC	32	36
Lighting	8	26
Special Functions		
Food Preparation	45	27
Food Storage	2	6
Sanitation	12	1
Other	1	4

Table 2-8. Energy Use in Schools

	Range (%)	Norms (%)
Environmental Control	45 to 80	65
Lighting and Wall Receptacles	10 to 20	15
Food Service	5 to 10	7
Hot Water	2 to 5	3
Special Functions	0 to 20	10

Table 2-9. Energy Use in Transportation Terminals

	Range (%)	Norms (%)
Space Heating	50 to 75	60
Lighting	5 to 25	15
Air Conditioning	5 to 25	15
Special Functions	3 to 20	10
Elevators, General Power, Parking, Security Lighting, Hot Water		

Table 2-10. Energy Use in Warehouses and Storage Facilities
(Vehicles Not Included)

	Range (%)	Norms (%) *
Space Heating	45 to 80	67
Air Conditioning	3 to 10	6
Lighting	4 to 12	7
Refrigeration	0 to 40	12
Special Functions	5 to 15	8
Elevators, General Power, Parking, Security Lighting, Hot Water		

* Norms for a warehouse or storage facility are strongly dependent on the products and their specific requirements for temperature and humidity control.

Table 2-11. Comparative Energy Use By System

		Heating & Ventilation	Cooling & Ventilation	Lighting	Power & Process	Domestic Hot Water
Schools	A	4	3	1	5	—
	B	1	4	2	5	3
	C	1	4	2	5	3
Colleges	A	5	2	1	4	3
	B	1	3	2	5	4
	C	1	5	2	4	3
Office Bldg.	A	3	1	2	4	5
	B	1	3	2	4	5
	C	1	3	2	4	5
Commercial Stores	A	3	1	2	4	5
	B	2	3	1	4	5
	C	1	3	2	4	5
Religious Bldg.	A	3	2	1	4	5
	B	1	3	2	4	5
	C	1	3	2	4	5
Hospitals	A	4	1	2	5	3
	B	1	3	4	5	2
	C	1	5	3	4	2

Climatic Zone A: Fewer than 2500 degree days
Climate Zone B: 2500—5500 degree days
Climate Zone C: 5500—9500 degree days

Source: Guidelines For Saving Energy In Existing Buildings ECM—1

Note: Numbers indicate energy consumption relative to each other
    (1) greatest consumption
    (5) least consumption

ENERGY ACCOUNTING

An important part of the overall energy auditing program is to be able to measure where you are, and determine where you are going. It is vital to establish an energy accounting system at the beginning of the program.

One of the more important aspects of energy management and conservation is measuring and accounting for energy consumption.

Most energy accounting systems have been devised and are administered by engineers for engineers. The engineers' principal interest in developing these systems has been the display of energy consumed per unit of production. That ratio has been called "energy efficiency," and changes in energy efficiency are clearly energy conserved or wasted. The engineer focuses all of his attention on reducing energy consumed per unit of production.

An energy efficiency ratio alone, however, cannot answer the kinds of questions asked by business managers and/or government authorities:

- If we are conserving energy, why is our total energy consumption increasing?
- If we are wasting energy, why is our total energy consumption decreasing?
- If we have made no change in energy efficiency, why is our energy consumption changing?
- How much of our energy consumption is due to factors beyond our control, such as weather, legislated environmental controls, etc.?
- How much of our energy consumption is the result of running experiments for saving energy?
- How much of our energy consumption is fixed variable?
- How much of our energy consumption change is due to increased production?

A VARIANCE ANALYSIS

Table 2-12 shows that only 1% more energy was used in 1991 but 14% more was paid for it than in 1990.

Table 2-12. Energy Accounting and Analysis

	1991	1990
Total Energy Used (Billions BTU)	6,847	6,804
Difference 1991/1990	43	
% Difference	1%	
Total Energy Cost ($000)	$19,015	$16,721
Difference 1991/1990	2,294	
$ Difference	14%	

These differences are caused by a number of factors. Those considered most important are changes in volume and/or mix of products produced, weather, pollution control requirements, alternative fuels (using more or less expensive fuels for the same application than those used in the base period), conservation, and price. All these factors or variances for analysis in both dollars and BTU should be isolated as shown in Table 2-13.

Table 2-13. Variance Analysis of Differences Between 1990 and 1991

	Variance	
Contributor	In Thousands of $	Billions of BTU
Volume/Mix	$1,054	418
Weather	(53)	(27)
Pollution	48	23
Other	403	142
Price	1,719	— —
Conservation	(877)	(513)
Alternative Fuels	88*	— —
Net Impact	$2,294	43

*Nonadditive Memo Item

Of the 43 billion BTU and $2,294,000 difference between 1990 and 1991, $1,054,000 and 418 billion BTU are the result of producing more products or a change in product mix. Should we want to separate volume from mix we can do it, but it is not really meaningful at the corporate level since only a handful of plants will have significant mix changes. We do it at the plant level if and when it is useful.

In metallurgy, the quality of the ore or whatever raw material is used can have a significant impact on energy used. The system can reflect this impact as well.

The weather was colder in the first quarter of 1991, but milder in the balance of the year, resulting in a reduced weather effect versus the prior year.

Forty-eight thousand dollars and 23 billion BTU were spent for additional pollution control in 1991 and 1990.

"Other" factors include such things as base loads which represent the fixed versus variable energy load or energy used for experimental processes or start-up of processes with no historical data for comparison. By putting such data into the "other" category, changes in energy efficiency in specific product lines can be isolated for analysis without being distorted by extraneous or extraordinary situations.

An additional $1,719,000 was paid for energy in 1991 because prices were up 13% overall.

Conservation avoided expenditures of $877,000 and 513 billion BTU. This was the result of a 7% reduction in total energy used per unit of production in 1991 versus 1990. Note that the conservation of 513 billion BTU is greater than the volume/mix increase of 418 billion BTU. *In effect, growth in demand was offset with conservation.*

Being forced to use more expensive alternative forms of energy during curtailments for other reasons cost $88,000 more in 1991. This item is not additive in summing the net impact as it is part of all other factors. It is tracked primarily for information and decision-making purposes, but not as a part of the variance analysis.

The net impact is the algebraic sum of all effects and is the total difference between energy used in this period and last. It is a closed loop which can be as complex or as simple as management wishes it to be. Whether a plant is a woodworking shop or a foundry, the plant manager has the option of making his analysis very deep, or as shallow as is required. In any case, all operations are additive to give a total corporate picture.

The analysis in dollars should be published each month as part of the corporate operations and financial progress report. The report is subtotaled by division, group and corporate total. Accompanying the numbers is a discussion of the current results and events, along with forecasts of price availability and any other factors subject to change.

Each reporting plant and division should receive a computer printout showing complete analysis by product line and by energy type or fuel, allowing plant managers and engineers to make decisions based on simple data analysis in whatever form they wish.

This system, plus conservation-oriented management, does increase energy efficiency.

What follows is a description of the Carborundum System followed by a discussion of analytical methods.

The Carborundum Energy Accounting and Analysis System should serve as a useful guide as to how one major energy intensive company established its accounting and analysis procedures.

ACCOUNTING INPUT

The data requirements can be as simple or as complex as the user chooses. The data form developed by Carborundum (Figure 2-4) is flexible enough to handle a broad spectrum of complexities.

FUEL USED

The first category of data required is the type of fuel or energy used and its cost. The quantity should be taken directly from utility invoices or inventory differences. One advantage of the Carborundum System is that the units may be metric or imperial or whatever the invoice shows.

Carborundum Energy Accounting and
Analysis System Data Input Form

Energy Management and Conservation Program

Plant _____

Division _____

Group _____

Today's Date _____

Plant Input Data

Period Covered _____

Description	Elec. kwh (000)*	Gas mcf	Oil gal. (000)*	Coal lbs. (000)*	Propane gal. (000)*	Other (000)*
Total Fuel Used						
Quantity						
Cost ($)						
** Conversion Factor	✕					
Production						
Product 1 NAME						
Production Unit						
Quant. Prod. (000)						
Fuel Used						
Product 2 NAME						
Production Unit						
Quant. Prod. (000)						
Fuel Used						
Product 3 NAME						
Production Unit						
Quant. Prod. (000)						
Fuel Used						
Product 4 NAME						
Production Unit						
Quant. Prod. (000)						
Fuel Used						
Product 5 NAME						
Production Unit						
Quant. Prod. (000)						
Fuel Used						
Heating						
Degree Days						
Fuel Used						
Cooling						
Degree Days						
Fuel Used						
Pollution Control						
Fuel Used						
Other						
Fuel Used						
** Alternate Fuel						

*All fuels reported in thousands to two decimal places.

Figure 2-4. Carborundum Energy Accounting and Analysis System Data Input Form

The Carborundum System works with five sources of energy: electricity, natural gas, fuel oil, coal and propane. No user need feel restricted to those five or feel that all of them need be used. An "other" category gives the system the flexibility for working with other fuels or forms of energy. In practice, we find that the "other" fuels are an extremely small proportion of our total. Our limited use of the "other" category has prevented distortions and has improved the quality of our analyses.

The conversion factor is the multiplier used to convert the unit of fuel to common thermal units. Carborundum uses British thermal units (BTU), but calories or joules or any other unit could be used.

For a simple energy accounting and analysis, only the fuel used and its cost are required. Each reporting location can stop there or give further data for more thorough analyses.

PRODUCTION

The first category of selection of the total energy distribution is production energy. The production quantity need not be in terms of units produced; it could be number of patrons of a movie theater, customers in a store, passengers in a transportation system, hours or other unit of time in which the facility is used, miles traveled, etc. In any case, the "product" should be clearly defined in the mind of the system's user and should be a common denominator of the business being conducted.

A product line may be a process as long as energy and production output are discretely measured.

The production unit should be in physical terms whenever possible—i.e. weight, volume, length, area, or number of items. Nonphysical units such as value, manhours, etc. are subject to distortions. Monetary value suffers from the distortion of inflation and other effects. Manhours worked is a poor energy accounting measure; there will be severe distortions with product mix or productivity changes. That is not to say that nonphysical units may not be used, but rather that the pitfalls of using them must be recognized.

To be useful, the amount of energy specifically used for production should be metered or at least allocated to production from the total. If production is low in energy intensity (energy cost as a per cent of production value), it is sometimes better to attribute all energy to heating and cooling, discussed below. If, on the other hand, energy used for heating and cooling is small compared to the energy used for production, all energy may be allocated to production.

The Carborundum System permits analysis of any number of different product lines for any one location or reporting unit. The input form allows for five product lines as that is the most usually reported. Plants wishing to report more than five may use the extra sheets or may break themselves down into subplants with separate input data for each subplant. Again, to be meaningful, the amount of energy specifically used for each product line must be measured independently or at least allocated from the total.

The amount of fuel or energy used for production is recorded in the same units for each fuel or energy form as the total at the top of the page. No conversion to a common thermal unit is made on the data input sheet.

HEATING AND COOLING

The next two categories of data requirement are heating and cooling. The reporting unit determines if these data are meaningful. The energy used should be discretely metered or allocated to heating and cooling. The amount used should be recorded in the same units for each fuel or energy forms as the total at the top of the input form.

The amount of comfort heating and cooling required is governed by weather conditions as opposed to process heating or cooling which should be considered production energy. For the purpose of this analysis, degree-days are used as the measure of weather intensity. Current and historical degree-day data are easily obtainable from airport or city weather departments, public utilities and other sources of fuel supply. Other factors such as wind, sun, exposure, etc. affect the heating and cooling energy requirements of buildings, but in our experience, degree-days are sufficiently significant and certainly simple enough.

This discussion will not include calculation of degree-days or other factors of weather intensity.

Other variables may affect the amount of heating and cooling energy used. One Carborundum plant varies its manufacturing area significantly as sales volume demands. The production use of energy is insignificant compared to that used for heating and cooling. To compensate for the varying conditions, the heating and cooling energy is treated as production energy and the unit of "production" is square feet of manufacturing space multiplied by degree-days. The use of the Carborundum System is limited only by the user's imagination.

POLLUTION CONTROL

One of the criticisms leveled at most energy reporting systems is the lack of recognition given to changing business conditions. Environmental control requirements are being met, but at the cost of greater energy expenditure to operate the pollution control devices. Many businessmen feel that energy requirements for pollution control diminish the impact of energy conservation efforts. The category of data recognizes pollution control energy requirements by considering it separately. Again, it is up to the reporting unit management to decide if the data are significant and if they will add to the usefulness of energy accounting and analysis. The energy must, of course, be measured or allocated from the total.

OTHER

The final category of data requirement is intended to give the system the flexibility of handling any other category considered useful by reporting unit management. We include in this category, energy consumed in experimental efforts, start-up of a new product line for which there is no history, and base loads of steam plants or process ovens and furnaces which are not variable and are therefore not affected by production volume or weather. Managers who do not want production and weather data analyses distorted by extraneous energy consumption will choose this category. The energy reported should be significant. This energy is not excluded from subsequent analysis, however.

ALTERNATE FUEL

The last line of the form (Figure 2-4) is to indicate those fuels used as alternates for each other anywhere in the reporting unit. In most cases it does not matter if some processes or energy uses have no alternative while others using the same fuel can be switched to another. The alternatives are recognized in the report form so long as they exist anywhere in the reporting unit. This subject will be covered more thoroughly in the discussion of output logic.

BALANCING

Once the input data are complete, one check should be made before treating it. The sum of fuel used for production, heating, cooling, pollution control and "other" should equal the total at the top of the sheet for each fuel used. If the totals do not balance, the error must be found before proceeding. The important thing here is that all energy is accounted for.

ACCOUNTING OUTPUT LOGIC

The output of the Carborundum energy accounting and analysis system is a display of treated data from two time periods and a comparison between them, identifying the nature and magnitude of any changes taking place. The selection of the two time periods is up to the user. The federal government compares current periods against the base year period of 1972. An industry association's reporting to the government would do the same, but for its own purposes it might prefer to compare this quarter vs. last quarter, of this year vs. last year, etc. A company or plant might want to compare this month vs. last month. The Carborundum Company prefers a monthly comparison of this year-to-date vs. last-year-to-date, and keeps a base period data on file for long-range comparisons as well.

The form of the output or display of data treated for analysis is not important but it should contain at least the elements of the Carborundum System output (Figure 2-5) discussed below.

The headings across the top of the form are fuels reported in the input. Each fuel has two columns; data from the current period and data from the period we are comparing against. The last pair of columns on the right are totals.

ENERGY USED "ENERGY (MBTU)"

The first line of data is the result of multiplying the amount of fuel used by the appropriate conversion factor and expressing the result in BTU. The conversion to BTU permits the summing of energy used by the five fuel types plus "other" fuels. The energy shown is the total used in the two time periods compared.

ENERGY COST AND COST PER UNIT OF ENERGY

Energy cost is the year-to-date sums of input data and the sum across the page for each of the two time periods in the comparison. Energy costs per unit are the result of dividing the costs by the energy used in BTU and expressing them in the form of $/BTU. Direct comparisons of the costs of two forms of energy can now be made, either in the same or different periods. The result of the exercise in the totals column is a weighted average cost per BTU. Shifts in the mix of fuels used will change the weighted average costs/BTU just as any change in individual fuel costs/BTU will affect the weighted averages. Opportunities are identified for overall energy cost reduction through fuel switching.

PRODUCTION

The next series of lines show production data including quantity produced, energy used for that production, energy efficiency, cost of energy used, energy saved (or wasted) as a result of changes in efficiency between time periods in the comparison and percent change in energy efficiency.

Energy used for production is expressed in BTU for each fuel type. Production quantity is not totaled for obvious reasons. Energy efficiency of production is expressed in BTU per thousand units and is the result of dividing the energy used for production by units produced.

The series of production data are repeated for as many product lines as the reporting unit management has chosen.

Obviously, the energy efficiency of different fuels and different product lines can now be compared in the same period or against efficiencies in other periods. Similarly, comparisons can be made of efficiency in one product line versus the efficiency in other periods in the same product line.

HEATING AND COOLING

The next six lines of data in the output show weather intensity in degree-days energy used for heating and cooling and energy efficiency of heating and cooling (energy used per degree-day) in the two time periods under consideration.

Weather intensity and its effects on energy used and energy efficiency can now be compared by fuel and between periods and in total.

POLLUTION CONTROL

The next line of data shows the energy used in BTU for pollution control in each period. If the requirements for pollution control are changing, the effects on energy consumption will be clearly shown.

OTHER

The last line of energy use data in the output shows the energy used in BTU in one period compared to another for such things as experimental efforts, start-up of a new product line for which there is no history and base loads of steam plants which are not weather. As described in the input section, this is a way of pulling these kinds of data out of other effects being

SEPTEMBER 1991 PLANT — MIDTOWN DIVISION — A B C GROUP — INDUSTRIAL COMPANY — U.S. INDUSTRY

DESCRIPTION	ELECTRICITY 91 SEP	ELECTRICITY 90 SEP	GAS 91 SEP	GAS 90 SEP	OIL 91 SEP	OIL 90 SEP	COAL 91 SEP	COAL 90 SEP	PROPANE 91 SEP	PROPANE 90 SEP	OTHER 91 SEP	OTHER 90 SEP	TOTAL 91 SEP	TOTAL 90 SEP
ENERGY (MBTU)	2408	2258	1625	2986	4210	5264	0	0	0	0	0	0	8243	10508
ENERGY ($)	15771	14215	3526	4807	10778	12212	0	0	0	0	0	0	30075	31234
COST/UT $/MB	6.55	6.30	2.17	1.61	2.56	2.32	0.00	0.00	0.00	0.00	0.00	0.00	3.65	2.97
PRODUCTION—														
PRODUCT LINE 1														
QUANTITY	1005	915	1005	915	1005	915	0	0	0	0	0	0		
ENERGY—MBTU	1417	1418	539	1860	1914	548	0	0	0	0	0	0	3870	3826
ENERGY/K-UT	1.410	1.550	.536	2.033	1.905	.599	0.000	0.000	0.000	0.000	0.000	0.000		
EN. COST ($)	9282	8930	1170	2994	4900	1271	0	0	0	0	0	0	15351	13195
EN. EFF—MBTU	141	0	1504	0-	1313	0	0	0	0	0	0	0	332	0
EN. EFF—o/o	9.049	0.000	73.617	0.000	******	0.000	0.000	0.000	0.000	0.000	0.000	0.000	7.911	0.000
PRODUCT LINE 2														
QUANTITY	31	28	52	47	50	200	0	0	0	0	0	0		
ENERGY—MBTU	763	700	989	939	1216	3628	0	0	0	0	0	0	2968	5267
ENERGY/K-UT	24.613	25.000	19.019	19.979	24.325	18.140	0.000	0.000	0.000	0.000	0.000	0.000		
EN. COST ($)	4998	4407	2146	1512	3113	8417	0	0	0	0	0	0	10257	14335
EN. EFF—MBTU	12	0	50	0	309	0	0	0	0	0	0	0	247	0
EN. EFF—o/o	1.550	0.000	4.803	0.000	34.100	0.000	0.000	0.000	0.000	0.000	0.000	0.000	9.091	0.000
PRODUCT LINE 3														
QUANTITY	150	100	0	0	0	0	0	0	0	0	0	0		
ENERGY—MBTU	51	35	0	0	0	0	0	0	0	0	0	0	51	35
ENERGY/K-UT	.340	.350	0.000	0.000	0.000	0.000	0.000	0.000	0.000	0.000	0.000	0.000		
EN. COST ($)	334	221	0	0	0	0	0	0	0	0	0	0	334	221
EN. EFF—MBTU	2	0	0	0	0	0	0	0	0	0	0	0	2	0
EN. EFF—o/o	2.924	0.000	0.000	0.000	0.000	0.000	0.000	0.000	0.000	0.000	0.000	0.000	2.924	0.000
HEATING—														
DEGREE DAYS	393	358	393	358	393	358	393	358	393	358	393	358		
ENERGY—MBTU	0	0	0	0	984	990	0	0	0	0	0	0	984	990
ENERGY/DD	0.000	0.000	0.000	0.000	2.504	2.764	0.000	0.000	0.000	0.000	0.000	0.000		
COOLING—														
DEGREE DAYS	10	44	10	44	10	44	10	44	10	44	10	44		
ENERGY—MBTU	15	67	0	0	0	0	0	0	0	0	0	0	15	67
ENERGY/DD	1.499	1.524	0.000	0.000	0.000	0.000	0.000	0.000	0.000	0.000	0.000	0.000		
POLLUTION CTL — ENERGY—MBTU	90	34	0	0	0	0	0	0	0	0	0	0	90	34
OTHER— ENERGY—MBTU	72	0	97	107	96	99	0	0	0	0	0	0	265	286

SEP 1990 TO SEP 1991 IMPACT —

VOLUME/MIX EFFECT (MBTU)							
PROD. LINE 1	140	183	54	0	0	—	376
PROD. LINE 2	75	100	2721	0	0	—	2546
PROD. LINE 3	18	0	0	0	0	—	18
TOTAL—MBTU	232	283	2667	0	0	—	2152
VOLUME/MIX EFFECT ($)							
PROD. LINE 1	878	295	125	0	0	—	1298
PROD. LINE 2	472	161	6312	0	0	—	5679
PROD. LINE 3	110	0	0	0	0	—	110
TOTAL—$	1461	455	6187	0	0	—	4271
WEATHER (MBTU)							
HEATING	0	0	97	0	0	—	97
COOLING	52	0	0	0	0	—	52
TOTAL	52	0	97	0	0	—	45
WEATHER ($)							
HEATING	0	0	224	0	0	—	224
COOLING	326	0	0	0	0	—	326
TOTAL	326	0	224	0	0	—	102
POLLUTION CONTROL—							
MBTU	56	0	0	0	0	—	56
$	352	0	0	0	0	—	352
OTHER—							
MBTU	72	90	3	0	0	—	21
$	452	145	6	0	0	—	300
ENERGY CONSERVATION—							
MBTU	158	1554	1519	0	0	—	193
$	996	2502	3525	0	0	—	27
ALTERNATE FUEL ADJUSTMENT—							
RATIO	0.00	.36	.64	0.00	0.00	—	0.00
DISTRIBUTION	0	2112	3723	0	0	—	0
ALTERNATE FUEL IMPACT—							
MBTU	0	487	487	0	0	—	
$	0	784	1130	0	0	—	346
PRICE ($)	613	910	1010	0	0	—	2534
NET IMPACT ($)—							
ACTUAL	1556	1281	1434	0	0	—	1159
PROGRAM	0	0	0	0	0	—	0
ENERGY CONSERVATION o/o							
ACTUAL	6.2	48.9	56.5	0.0	0.0	—	2.3
PROGRAM	0.0	0.0	0.0	0.0	0.0	—	0.0

Figure 2-5. Energy Management and Conservation Report

isolated for analysis. At the same time, the energy is not excluded from the overall analysis—no energy is excluded.

THIS PERIOD VS. LAST PERIOD IMPACT

Beyond this point in our output, the numbers define the variance analysis which show the impact of changing volume, product mix, weather, pollution control, energy conservation efforts, "Other" effects, and price, on energy usage and cost in the period being analyzed compared to the period used as the basis for analysis. The algebraic sum of all the impacts or variances is the net impact. If it is a valid variance analysis, the net impact is the difference between the total energy used or cost in the period being analyzed and that in the base period. If the net impact is not equal to the difference between totals, there is an error in arithmetic or assumptions.

Also shown among impacts is the effect of using alternative fuels either because of curtailments or cost. This variance is not additive in the variance analysis since the effect cannot be isolated from the other effects without considerable arithmetic. We calculate the impact of alternative fuels only to display it as an economic fact to be taken into consideration for management strategic and decision-making purposes. The energy impact is zero in this system since alternatives are used on a straight BTU exchange basis.

This is not strictly true, of course, as calorific values are not always useful energy values and there are often differences in efficiency of combustion or use. These effects are usually small enough to ignore. If a user feels that they are significant, there is no reason why they cannot be taken into consideration.

VOLUME/MIX IMPACT

The impact of volume and/or product mix changes is the amount of more (or less) energy that is used currently, as opposed to previously, solely as the result of producing more (or less) product or proportionately more (or less) energy intense products.

In mathematical terms, the effect of volume and/or product mix changes on energy is defined as the algebraic sum of the products of the change in volume in the periods under discussion in each product line multiplied by the energy efficiency of production of each product line in the base period. In the Carborundum System this translates into the algebraic sum of the products of the difference in production volume this year-to-date and last year-to-date times the energy used per unit of production last year-to-date for each product line. The monetary impact of volume/mix is the energy impact calculated as above multiplied by the cost per unit of energy last year-to-date. That is to say, the impact of changes in volume/mix on energy use or cost is the difference between this period's volume mix and last period's volume mix times the energy efficiency achieved in the last or base period. The result ignores improvements in efficiency (identified later as energy conservation effects) and inflation (identified later as price effects) and isolates the effects of only volume/mix changes. A more detailed treatment of this effect appears in the analysis section.

WEATHER IMPACT

The effect of weather changes (colder winter or hotter summer) on energy consumption is defined as the change in degree-days in the periods under discussion times the heating or cooling efficiency in the period used as the basis for analysis. In the Carborundum System, this translates into the difference in degree-days this year-to-date and last year-to-date times the energy used per degree-day last year-to-date. The monetary impact of weather is the impact calculated as above times the cost per unit of energy last year-to-date. That is to say, the impact of weather changes on energy use or cost is the difference between this period's weather and last, times the heating/cooling energy efficiency in the last or base period. The result ignores improvements in efficiency (identified later as energy conservation effects) and inflation (identified later as price effects), and isolates the effect of weather.

POLLUTION CONTROL IMPACT

The impact of the energy increase or decrease to control pollution in the current period versus any other time period is simply the difference in the energy used in the two periods. The financial impact is the impact calculated above multiplied by the cost per unit of energy in the last period. The result ignores conservation and price effects as before, and isolates the effect of pollution control.

"OTHER" IMPACTS

The impact of other energy uses, previously defined as experimental, start-up of product lines without history, of base loads, etc., is simply the difference in energy used in the two periods being compared. The economic impact is the impact calculated above multiplied by the cost per unit of energy in the prior period. Again, the result ignores conservation and price effects and isolates the effect of these "other" uses of energy.

ENERGY CONSERVATION

In the Carborundum System, conservation is defined as the reduction in energy consumed in the current period versus that consumed in the comparative period after all other effects (volume/mix, weather, pollution, "other" effects) have been taken into consideration. To arrive at the energy conserved (in BTU) the algebraic sum of all the impacts calculated above is added to the total used in the prior period and that sum subtracted from the total used in the current period. To check the calculation, the sum of all the calculated energy savings in each product line and/or heating and cooling should equal the number derived by subtraction above.

The financial impact of conservation is the product of all the energy conserved multiplied by the costs per unit in the prior period. Some people are concerned about "error" arising from using prior period vs. current period energy unit costs.

The magnitude of "error" is not large and if objectives or goals are stated in the same dollars, there should be no problem. In any case, the correction may be made for discussion purposes. Senior management seems to prefer conservation stated in last year's dollars because in the goal-setting process, those are "current dollars" at the time goals are set.

Objective achievement in conservation is thus not distorted by price changes which are more difficult to predict. When comparing performance to a more distant base period, the "error" of using the prior period dollars becomes significant and must be adjusted for. In any case, expressing the conservation impact in the last year's dollars, isolates conservation from price effects.

PRICE IMPACT

The impact of price changes is simply the difference in unit cost in the current period and those in the comparative period multiplied by the energy used in the current period.

NET IMPACT

The net impact, then, is the net effect of all the isolated factors identified. The system identifies only the economic net impact in the output format. There is, of course, an energy net impact (in BTU) which would be useful primarily to suppliers, but not really to a user unless he has a limited captive supply. It is certainly of interest to the government.

In either case, the net impact is the algebraic sum of all the isolated factors minus the conservation effect. By definition, the net impact is the difference between the total energy used or its cost in the current period and that of the period being used as a basis for the analysis. In the system output format, the differences referred to are those between fuel or total columns in the first two lines of the output. If the net impact by the sum-of-effects method is not equal to the difference between the period and last, there is an error in either the data or the calculations.

Every plant manager, division manager and group vice-president should be committed to a "net impact" for the coming year. This number is tracked versus performance in the output.

ENERGY CONSERVATION PERCENT

The last two lines of the energy accounting and analysis system output are the percent energy conserved, actual and programmed. This is an overall conservation percentage change in the current period vs. the prior period. The percentage is arrived at by dividing energy conserved by energy conserved plus the total energy used in the current period.

Commitment by all levels of management at Carborundum to a conservation percentage is made prior to the start of the year and performance is measured against it.

ANALYSIS

An example of a plant accounting and analysis for 1991 vs. 1990 September year-to-date is shown in Figure 2-5. For the most part, data are taken from a combination of real plant situations to illustrate the simplicity and flexibility of the system.

INPUT

Fuel usage data are taken directly from invoices and/or meters. Production data are readily available. Notice the variety of physical units used. Production Line 1 is plastic products and the unit (lbs) is constant for all fuel forms as is quantity produced. In glass products, Product Line 2, the unit is number of items produced with electricity and oil, but the unit is pounds for those produced with gas. Different numbers of items were produced electrically—and oil-fired in the same month. This is not an unusual situation. Gas process production has traditionally been measured in weight of throughput. accounting and analysis. Product Line 3, wooden products, requires only electricity.

Plant management could just as easily not have broken products into three lines, but by doing so, we will more properly see the effect of production mix/volume changes. Most energy is metered because we have found that the simple act of metering results in energy awareness which leads to conservation.

September usually has heating degree-days (cold days) and cooling degree-days (warm days). Therefore, both heating and cooling processes took place. It is seen that September 1991 was cooler than September 1990.

More pollution control equipment was installed in the year between Septembers increasing energy consumption. Plant management chose to study the effect through input of pollution control data.

The plant has three loads of the "other" variety. There are experiments under way to replace gas with electricity in furnacing. Oil-fired equipment has a fixed load which does not vary with production levels. The oil-fired boiler has a fixed load which is unaffected by weather or production.

Since gas and oil are alternate fuels for the plastics line, they are checked in the input data sheet to alert the analyst. It does not matter how much or where the alternates are used for the output logic to be run on the data.

OUTPUT

Using the input data and completing the arithmetic described in the output logic section (Figure 2-6), we produced the numbers shown in Figure 2-5. Analysis results in the following conclusions starting from the top of Figure 2-5:

1. A 20% drop in total energy use only resulted in a 5% drop in total energy cost. Gas and oil were the major contributors to usage drops and to cost increase.
2. Gas prices ($/MBTU) are up 35%, oil up 10% and electricity up barely 4%. The weighted average energy unit cost is up 22%. If the mix of energy forms had been the same this year as last, the weighted average would only have increased 10%.

ARITHMETIC FORMULAE USED IN THE CARBORUNDUM ACCOUNTING AND ANALYSIS SYSTEM

The numbers (#) used in the following equations refer to the item number at the left margin. MBTU is mega or millions of BTU. YTD is this year to date or this period. LYTD is last year to date or prior period; the period against which comparisons are being made.

1. Energy Used (MBTU) = Invoice or equivalent quantity X input conversion factor to MBTU
2. Energy Dollars = Invoice or equivalent dollars
3. Cost/Unit ($/MBTU) = #2 ÷ #1
4. Production Quantity = Quantity produced from input sheet
5. Production Energy (MBTU) = Input value in invoiced units X input conversion factor to MBTU
6. Energy (MBTU) /Unit of production = #5 ÷ #4
7. Production Energy Cost ($) = #5 X #3
8. Production Energy Saved (MBTU) = (#4 YTD X #6 LYTD) − #5 YTD
9. Production Energy Conservation (%) = (#8 ÷ (#8 + #5 YTD)) X 100
10. Heating Degree Days = Input value
11. Heating Energy (MBTU) = Input value in invoiced units X input conversion factor to MBTU
12. Heating Energy/Degree-Day (MBTU/DD) = #11 ÷ #10
13. Heating Energy Saved (MBTU) = (#10 YTD X #12 LYTD) − #11 YTD
14. Cooling Degree Days = Input value
15. Colling Energy (MBTU) = Input value in invoiced units X input conversion factor to MBTU
16. Cooling Energy/Degree-Day (MBTU/DD) = #15 ÷ #14
17. Cooling Energy Saved (MBTU) = (#14 YTD X #16 LYTD) − #15 YTD
18. Pollution = Input value X input conversion factor to MBTU
19. Other = Input value in invoiced units X input conversion factor to MBTU

A useful check at this point is that the sum of all energy used for production, heating, cooling, pollution and "other" (items 5, 11,

Figure 2-6. Energy Accounting Instructions

15, 18, 19) must equal the total energy used (item 1).

Impacts or Variances

The following is a variance analysis to account for the difference in gross energy used or cost. Cost variance is in terms of last year's dollars, except for price.

20. Volume/Mix effect, Product line 1 (MBTU) = (#4 YTD − #4 LYTD) X #6 LYTD
21. Volume/Mix effect, Product line 1 ($) = #20 X #3 LYTD
22. Weather effect (MBTU) heating = (#10 YTD − #10 LYTD) X #12 LYTD
23. Weather effect (MBTU) cooling = (#14 YTD − #14 LYTD) X #16 LYTD
24. Weather effect ($) heating = #22 X #3 LYTD
25. Weather effect ($) cooling = #23 X #3 LYTD
26. Pollution Control effect (MBTU) = #18 YTD − #18 LYTD
27. Pollution Control effect ($) = #26 X #3 LYTD
28. Other effects (MBTU) = #19 YTD − #19 LYTD
29. Other effects ($) = #28 X #3 LYTD
30. Energy Conservation (MBTU) = #1 LYTD + #20 + #22 + #23 + #26 + #28 − #1 YTD

Energy conservation can be checked by adding items 8, 13, and 17 which must equal item 30.

31. Energy Conservation ($) = #30 X #3 LYTD
32. Alternate fuel ratio = Each alternate fuel LYTD ÷ Total of alternate fuels LYTD
33. Alternate fuel distribution = #32 X Total of alternate fuels YTD
34. Alternate fuel impact (MBTU) = #1 YTD − #33
35. Alternate fuel impact ($) = #34 X #3 LYTD
36. Price ($) = (#3 YTD − #3 LYTD) X #1 YTD
37. Net Impact (MBTU) = #20 + #22 + #23 + #26 + #28 − #30
38. Net Impact ($) = #21 + #24 + #25 + #27 + #29 − #31 + #36

Net impacts can be checked by subtracting item 1, LYTD from item 1, YTD in MBTU and item 2 LYTD from item 2 YTD in dollars which must equal items #37 and #38 respectively.

39. Energy Conservation % = (#30 ÷ (#30 + #1 YTD)) X 100

Figure 2-6. Concluded

3. In spite of the bargain that gas prices still represent over oil (30% cheaper last year, 15% cheaper now), the price gap is closing fast and gas curtailment is forcing us to use proportionately more oil.

4. A small amount of electricity saved is worth a great deal more than the same amount of gas or oil because of the 3:1 cost ratio.

5. Production energy efficiency in plastic products (Product Line 1) is 9.049% improved in electric. Since gas and oil are alternates, the sum of the energy efficiencies for gas and oil for each period divided by each other will yield energy efficiency improvement of the process of 7.257%. The combined effect can be seen in the "total" column where the efficiency improvement is 7.911% for the plastics line and the total saved is 322 MBTU or about 4% of the total energy of the plant. Since the total energy saved is 193 MBTU, the plastics line is clearly the hero of the plant having saved 332 MBTU.

   To calculate MBTU saved by production efficiency, multiply production quantity this period by energy efficiency last period and subtract energy used this period.

   Notice that electricity was saved through efficiencies while gas was saved because of a switch to oil.

6. Increase in production volume of plastic helped conservation efforts but cannot account for it all. The increase in volume was only about 10% vs. the total energy efficiency gain of 7.911%; therefore, whatever effort was made was successful and should be encouraged to continue.

7. Fuel curtailment forced Carborundum to switch to an alternate, which affected the line's use of energy significantly.

8. Glass products production was up 10% in electric and gas, but off 75% in oil. The 1.550% efficiency improvement in electric and a 4.803% improvement in gas was shattered by the 34.100% loss of efficiency in oil, which was due to loss of production volume in '91. Apparently September '90 production with oil was a special run be-

cause of the disparity of unit production between electric and oil. The result is a loss of 247 MBTU.

The possibility of inventorying and batch processing glass products is being explored in the oil-fired process, since experience of September '90 indicates large potential efficiency gains.

9. Almost half of the energy dollars is spent on the plastic line so our conservation effort is properly directed; that is where we have saved the most. Energy intensity (MBTU/K Unit) is highest, however, in the glass line where a third of our energy dollars is spent. More conservation effort in the glass line would have a greater impact on profit improvement.

10. Wooden products have little energy impact and the insignificant energy improvement shows that little effort has been spent on improving efficiency. In view of the potential impact, this is probably as it should be, but it does show a lack of an energy awareness which would impact all energy use more significantly.

11. In spite of September '91 being 10% colder than September '90, energy used for heating was virtually the same because thermostats were kept down. The energy saved is second only to the 332 MBTU saved in the plastics line which would indicate the desirability of devoting more effort towards improving insulation and other space heating conservation devices.

12. The requirement for cooling energy was greatly reduced but no significant cooling energy was conserved. Cooling is done by electricity, however, which is two and one-half times the cost of heating oil. Improvements in cooling will yield proportionately more for the dollar than improvements in oil heating.

13. A 10% saving in heating and cooling energy is low. Our experience is that 30% conservation is usually possible in systems over five years old. A 30% improvement would be from ground zero, however, and this output is from last year. Prior conservation efforts are not shown but can be calculated to test for more potential savings.

14. Pollution control energy is a much smaller proportion of the total than most managements believe. That does not mean, however, that its efficiency cannot be improved upon, just that it won't yield very much in savings in proportion to the total plant. In most cases, pollution control devices are electric which is the most expensive energy and from a return on investment point of view, it often is a good savings potential. Maintenance is often all it takes.

15. The use of electricity in the "OTHER" category is for experimental purposes which were not done in 1990. Should the experiments prove fruitful and permanent equipment be installed, the category would be removed from "other" and added to production either in an existing product or process or in a new one as appropriate.

   The use of gas under "other" is the fixed load of two tunnel kilns. One was not fired this year; hence the reduction. If the fixed load is reduced through additional insulation it would be reflected.

   The use of oil under "other" is the fixed load of a boiler and the same comment applies as with tunnel kilns.

In the "YTD-90 to YTD-91 IMPACT" section the interpretation of the numbers is: "How much more or less was used as a result of the variance identified?"

16. The volume/mix effect shows that the plastics line used a total of 376 MBTU more in '91 than in '90 and paid $1298 more for energy only because of the increase in volume. The glass line volume decrease in the oil process cost $5679 less and the loss of efficiency increased the cost by $716.88 (−309 MBTU saved × $2.32/MBTU '90-YTD).

17. The fact that the weather was cooler increased heating costs by $224 but decreased cooling costs by $326 yielding a net reduction in cost of $102 emphasizing the leverage of improving cooling efficiency.

18. The impact of "other" is a reduction in energy (−21 MBTU) but an increase in cost ($300) primarily because

of the experimental work in electricity and in spite of the fact that a tunnel kiln was shut down.

19. A most important effect is the fact that energy was conserved (193 MBTU) but dollars were wasted (−$27 saved)! Partly due to the requirement to switch to oil but mostly because of the loss in volume in the oil-fired glass process. Possibly a scheduling problem which could be avoided.

20. The impact of having to switch to oil was $346 which, in view of the total energy cost is not much but a cost nevertheless.

21. Price increases cost more than any other impact other than the loss of volume in the oil-fired glass process.

22. The net impact of the energy situation was a reduction in the total cost of energy of $1159. One conclusion here is that it was pretty cheap energy which was not used. Clearly planning is called for.

23. The net 2.3% conservation is not to be proud of in our experience.

There are other conclusions to be drawn which require standard financial report numbers. Management decisions/ actions based on those already outlined above would include such things as:

- Seek more ways to conserve electricity.
- Find methods of improving efficiency of use of oil in the plastic products line.
- Explore all conservation possibilities in the glass product line.
- Determine economic consequences of inventorying and batch-processing of the oil-fired glass products versus continuous burning.
- Look for more savings potential in the heating and cooling systems.
- Develop an energy awareness in the labor force and at all levels of management.

SUMMARIES

The next level of energy accounting and analysis is the aggregation of data outputs from other locations into division and/or company summaries. Beyond that there can be industry summaries (perhaps by industry associations) ending up finally, in national summaries.

Every number in the right-hand totals column of the output (except cost per BTU and percent conservation) can be added to those in any other output regardless of complexity. Cost per BTU and percent conservation are then calculated from grand totals. Degree-days are not additive as the total would be meaningless. No production quantities are carried forward in combining with other output. To show production quantities in summaries would require an endless number of product lines, needlessly complicating the analysis. Industry production totals would be entered on industry input data forms similar to plant input forms by participants in any industrial energy reporting system. Analysis by product line is carried out at plant levels and summarized verbally at higher levels.

The potential for justification numbers for capital investment should be obvious.

AND FINALLY . . .

Experience at Carborundum has been that a location achieves energy conservation and profit improvement very soon after the installation of this system. Whether locations are factories, warehouses, shipping operations or offices, energy and money are saved.

3

Survey Instrumentation

To accomplish an energy audit survey it is necessary to clarify energy uses and losses. This chapter illustrates various types of instruments which can aid in the energy audit survey.

MEASURING BUILDING LOSSES

Infrared energy is an invisible part of the electromagnetic spectrum. It exists naturally and can be measured by remote heat-sensing equipment. Within the last four years lightweight portable infrared systems became available to help determine energy losses. Differences in the infrared emissions from the surface of objects cause color variations to appear on the scanner. The hotter the object, the more infrared radiated. With the aid of an isotherm circuit the intensity of these radiation levels can be accurately measured and quantified. In essence the infrared scanning device is a diagnostic tool which can be used to determine building heat losses. Equipment costs range from $400 to $25,000.

An overview energy scan of the plant can be made through an aerial survey using infrared equipment. Several companies offer aerial scan services. Aerial scans can determine underground stream pipe leaks, hot gas discharges, leaks, etc.

Since IR detection and measurement equipment have gained increased importance in the energy audit process, a summary of the fundamentals are reviewed in this section.

In addition to detecting building energy losses IR Thermography has been used for other applications, listed in Table 3-1.

Table 3-1. Applications of IR Thermography

Inspection of power transmission equipment.
Water leakage into building roof insulation.
Checking for poor building insulation.
Detection of thermal pollution in rivers and lakes.
Studying coating uniformity on webs.
Inspecting cooling coils for plugged tubes.
Medical uses involving early detection of malignant tumors.
Spotting plugs and air locks in condenser tubes.
Controlling paper calendaring operations.
Studying the behavior of thermal sealing equipment.
Investigating ultrasonic sealers and sealing operations.
Inspection of electronic circuits.
Hot injection molding problems.
Studying the behavior of heating and cooling devices.
Detection of plugged furnace tubes.
Examination of consumer products for hot spots.
Spotting defects in laminated materials.
Finding leaks in buried steam lines.
Inspection of heavy machinery bearings.
Study of stresses due to thermal gradients in a component.
Detection of defects such as voids and inclusions in castings.

INFRARED RADIATION AND ITS MEASUREMENT

The electromagnetic spectrum is illustrated in Figure 3-1.

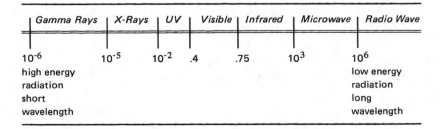

Figure 3-1. Electromagnetic Spectrum

The visible portion of the spectrum runs from .4 to .75 micrometers (μm). The infrared or thermal radiation begins at this point and extends to approximately 1000 μm. Objects such as people, plants, or buildings will emit radiation with wavelengths around 10 μm.

Infrared instruments are required to detect and measure the thermal radiation. To calibrate the instrument a special "black body" radiator is used. A black body radiator absorbs all the radiation that impinges on it and has an absorbing efficiency or emissivity of 1.

The accuracy of temperature measurements by infrared instruments depends on the three processes which are responsible for an object acting like a black body. These processes—absorbed, reflected, and transmitted radiation—are responsible for the total radiation reaching an infrared scanner.

The real temperature of the object is dependent only upon its emitted radiation.

Corrections to apparent temperatures are made by knowing the emissivity of an object at a specified temperature.

The heart of the infrared instrument is the infrared detector. The detector absorbs infrared energy and converts it into electrical voltage or current. The two principal types of detectors are the thermal and photo type. The thermal detector generally requires a given period of time to develop an image on photographic film. The photo detectors are more sensitive and have a higher response time. Television-like displays on a cathode ray tube permit studies of dynamic thermal events on moving objects in real time.

There are various ways of displaying signals produced by infrared detectors. One way is by use of an isotherm contour. The lightest areas of the picture represent the warmest areas of the subject and the darkest areas represent the coolest portions. These instruments can show thermal variations of less than .1°C and can cover a range of −30°C to over 2000°C.

The isotherm can be calibrated by means of a black body radiator so that a specific temperature is known. The scanner can then be moved and the temperatures of the various parts of the subject can be made.

There are many applications of infrared scanning devices as illustrated in Table 3-1.

Figure 3-2 illustrates the use of an aerial thermogram to detect heat losses. The information contained in this section should not be construed as a recommendation or complete listing. It simply serves as a sample of the products available.

MEASURING ELECTRICAL SYSTEM PERFORMANCE

The ammeter, voltmeter, wattmeter, power factor meter and footcandle meter are usually required to do an electrical survey. These instruments are described below.

AMMETER AND VOLTMETER

To measure electrical currents, ammeters are used. For most audits alternating currents are measured. Ammeters used in audits are portable and are designed to be easily attached and removed.

Figure 3-3 illustrates two typical types of meters which can be used for current or voltage measurements. Notice the meter illustrated on page 49 can be clamped around the conductors to measure current.

There are many brands and styles of snap-on ammeters commonly available that can read up to 1000 amperes continuously. This range can be extended to 4000 amperes continuously for some models with an accessory step-down current transformer.

The snap-on ammeters can be either indicating or recording with a printout. After attachment, the recording ammeter can keep recording current variations for as long as a full month on one roll of recording paper. This allows studying current variations in a conductor for extended periods without constant operator attention.

The ammeter supplies a direct measurement of electrical current which is one of the parameters needed to calculate electrical energy. The second parameter required to calculate energy is voltage, and it is measured by a voltmeter.

Several types of electrical meters can read the voltage or

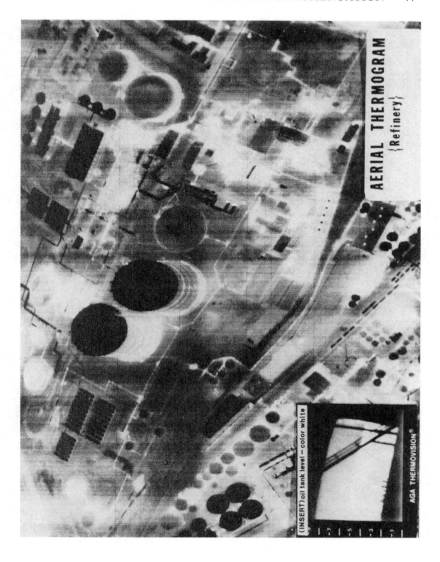

Figure 3-2.
Aerial
Thermogram
(Photograph
courtesy
AGA Co.
Secausus,
New Jersey)

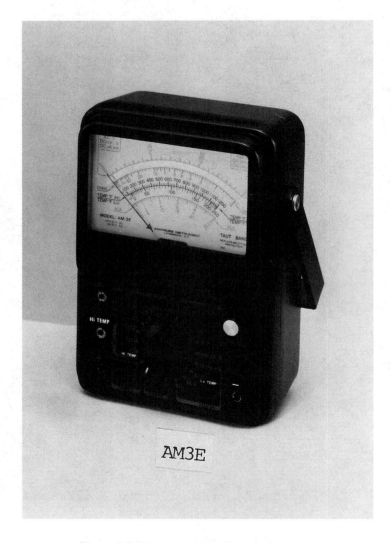

Figure 3-3. Meters Used in Electrical Surveys
*(Photographs courtesy of Amprobe Instrument,
Division of Core Industries, Inc.)*

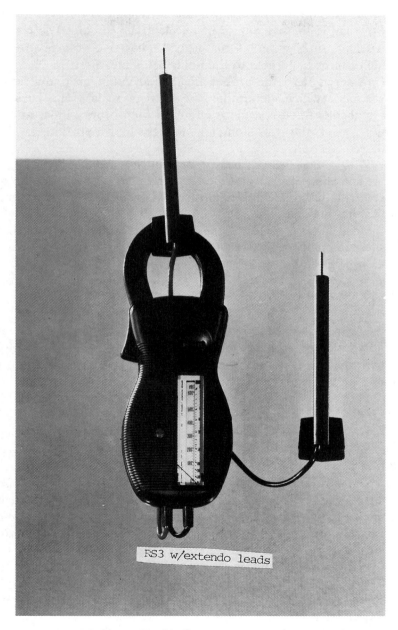

Figure 3-3. Meters Used in Electrical Surveys (concluded)

current. A voltmeter measures the difference in electrical potential between two points in an electrical circuit. The meter illustrated in Figure 3-3 on page 49 has two voltage probes which can be hand held to the point being measured.

In series with the probes are the galvanometer and a fixed resistance (which determine the voltage scale). The current through this fixed resistance circuit is then proportional to the voltage and the galvanometer deflects in proportion to the voltage.

The voltage drops measured in many instances are fairly constant and need only be performed once. If there are appreciable fluctuations, additional readings or the use of a recording voltmeter may be indicated.

Most voltages measured in practice are under 600 volts and there are many portable voltmeter/ammeter clamp-ons available for this and lower ranges.

WATTMETER AND POWER FACTOR METER

The portable wattmeter can be used to indicate by direct reading electrical energy in watts. It can also be calculated by measuring voltage, current and the angle between them (power factor angle).

The basic wattmeter consists of three voltage probes and a snap-on current coil which feeds the wattmeter movement.

The typical operating limits are 300 kilowatts, 650 volts, and 600 amperes. It can be used on both one- and three-phase circuits.

The portable power factor meter is primarily a three-phase instrument. One of its three voltage probes is attached to each conductor phase and a snap-on jaw is placed about one of the phases. By disconnecting the wattmeter circuitry, it will directly read the power factor of the circuit to which it is attached.

It can measure power factor over a range of 1.0 leading to 1.0 lagging with "ampacities" up to 1500 amperes at 600 volts. This range covers the large bulk of the applications found in light industry and commerce.

The power factor is a basic parameter whose value must be

known to calculate electric energy usage. Diagnostically it is a useful instrument to determine the sources of poor power factor in a facility.

Digital read-outs of energy usage in both KWH and KW Demand or in Dollars and Cents, including Instantaneous Usage, Accumulated Usage, Projected Usage for a particular billing period, Alarms when over-target levels desired for usage, and Control-Outputs for load-shedding and cycling are possible.

Continuous displays or intermittent alternating displays are available at the touch of a button of any information needed such as the cost of operating a production machine for one shift, one hour or one week.

A typical power meter is illustrated in Fig. 3-4.

FOOTCANDLE METER

Footcandle meters measure illumination in units of footcandles through light-sensitive barrier layer of cells contained within them. They are usually pocket size and portable and are meant to be used as field instruments to survey levels of illumination. Footcandle meters differ from conventional photographic lightmeters in that they are color and cosine corrected.

TEMPERATURE MEASUREMENTS

To maximize system performance, knowledge of the temperature of a fluid, surface, etc. is essential. Several types of temperature devices are described in this section.

THERMOMETER

There are many types of thermometers that can be used in an energy audit. The choice of what to use is usually dictated by cost, durability, and application. Figure 3-5 illustrates a common type of temperature measuring device.

For air-conditioning, ventilation and hot-water service applications (temperature ranges 50°F to 250°F) a multipurpose portable battery-operated thermometer is used. Three separate

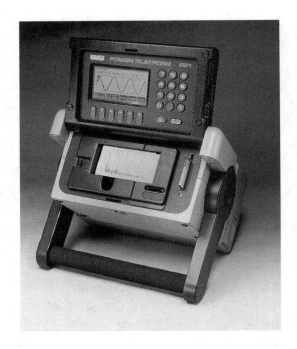

Figure 3-4. Power Meter
(Photograph courtesy of Dranetz Technologies, Inc.)

probes are usually provided to measure liquid, air or surface temperatures.

For boiler and oven stacks (1000°F) a dial thermometer is used. Thermocouples are used for measurements above 1000°F.

SURFACE PYROMETER

Surface pyrometers are instruments which measure the temperature of surfaces. They are somewhat more complex than other temperature instruments because their probe must make intimate contact with the surface being measured.

Surface pyrometers are of immense help in assessing heat losses through walls and also for testing steam traps.

They may be divided into two classes: low-temperature (up to 250°F) and high-temperature (up to 600-700°F). The low-

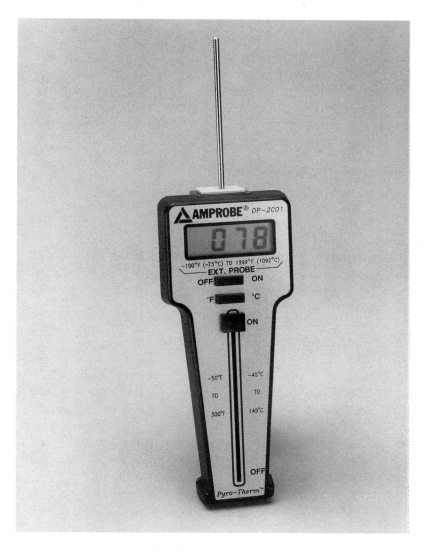

Figure 3-5. Thermometer
(Photograph courtesy of Amprobe Instrument, Division of Core Industries, Inc.)

temperature unit is usually part of the multipurpose thermometer kit. The high-temperature unit is more specialized, but needed for evaluating fired units and general steam service.

There are also noncontact surface pyrometers which measure infrared radiation from surfaces in terms of temperature. These are suitable for general work and also for measuring surfaces which are visually but not physically accessible.

A more specialized instrument is the optical pyrometer. This is for high-temperature work (above 1500°F) because it measures the temperature of bodies which are incandescent because of their temperature.

PSYCHROMETER

A psychrometer is an instrument which measures relative humidity based on the relation of the dry-bulb temperature and the wet-bulb temperature.

Relative humidity is of prime importance in HVAC and drying operations. Recording psychrometers are also available. Above 200°F humidity studies constitute a specialized field of endeavor.

PORTABLE ELECTRONIC THERMOMETER

The portable electronic thermometer is an adaptable temperature measurement tool. The battery-powered basic instrument, when housed in a carrying case, is suitable for laboratory or industrial use.

A pocket-size digital, battery-operated thermometer is especially convenient for spot checks or where a number of rapid readings of process temperatures need to be taken.

THERMOCOUPLE PROBE

No matter what sort of indicating instrument is employed, the thermocouple used should be carefully selected to match the application and properly positioned if a representative temperature is to be measured. The same care is needed for all sensing devices—thermocouple, bimetals, resistance elements, fluid expansion and vapour pressure bulbs.

SUCTION PYROMETER

Errors arise if a normal sheathed thermocouple is used to measure gas temperatures, especially high ones. The suction pyrometer overcomes these by shielding the thermocouple from wall radiation and drawing gases over it at high velocity to ensure good convective heat transfer. The thermocouple thus produces a reading which approaches the true temperature at the sampling point rather than a temperature between that of the walls and the gases.

MEASURING COMBUSTION SYSTEMS

To maximize combustion efficiency it is necessary to know the composition of the flue gas. By obtaining a good air-fuel ratio substantial energy will be saved.

COMBUSTION TESTER

Combustion testing consists of determining the concentrations of the products of combustion in a stack gas. The products of combustion usually considered are carbon dioxide and carbon monoxide. Oxygen is tested to assure proper excess air levels.

The definitive test for these constituents is an Orsat apparatus. This test consists of taking a measured volume of stack gas and measuring successive volumes after intimate contact with selective absorbing solutions. The reduction in volume after each absorption is the measure of each constituent.

The Orsat has a number of disadvantages. The main ones are that it requires considerable time to set up and use and its operator must have a good degree of dexterity and be in constant practice.

Instead of an Orsat, there are portable and easy to use absorbing instruments which can easily determine the concentrations of the constituents of interest on an individual basis.

Setup and operating times are minimal and just about anyone can learn to use them.

The typical range of concentrations are CO_2: 0–20%, O_2: 0–21% and CO: 0–0.5%. The CO_2 or O_2 content along with knowledge of flue gas temperature and fuel type allow the flue gas loss to be determined off standard charts.

BOILER TEST KIT

The boiler test kit illustrated in Figures 3-6, 3-7, 3-8, and 3-9 contains the following:

CO_2 Gas Analyzer,
O_2 Gas Analyzer,
   Inclined Monometer,
CO   Gas Analyzer.

The purpose of the components of the kit is to help evaluate fireside boiler operation. Good combustion usually means high carbon dioxide (CO_2), low oxygen (O_2), and little or no trace of carbon monoxide (CO).

GAS ANALYZERS

The gas analyzer illustrated in Figure 3-6 is the Fyrite type. The Fyrite type differs from the Orsat apparatus in that it is more limited in application and less accurate. The chief advantages of the Fyrite are that it is simple and easy to use and is inexpensive. This device is many times used in an energy audit. Three readings using the Fyrite analyzer should be made and the results averaged.

DRAFT GAUGE

The draft gauge is used to measure pressure. It can be the pocket type shown in Figure 3-9, or the inclined monometer type shown with the test kit.

Gas Analyzer

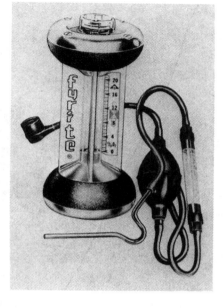

Carbon Monoxide Analyzer

Figure 3-6. Boiler Test Kit

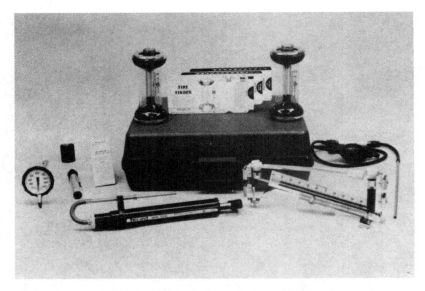

Test Kit *(above)* and Smoke Tester *(below)*

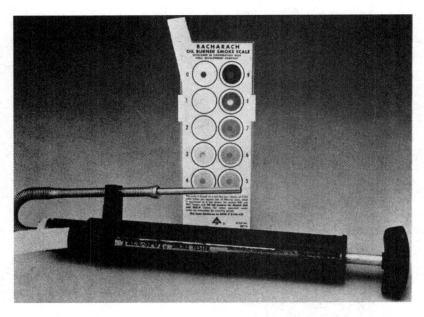

(Photographs these two pages courtesy of Bacharach Instrument Company)

Figure 3-7. Boiler Test Kit

Air Filter Gauge

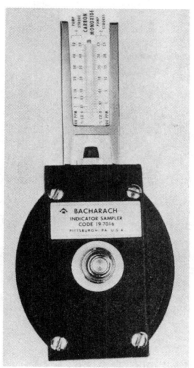

Hydrogen Sulfide Detector

**Figure 3-8. Components for
Combustion Testing**
*(Photographs courtesy of
Bacharach Instrument Company)*

Draft Gauge

(Photo courtesy of Bacharach Instrument Company)

Digital Sling Psychrometer

(Photo courtesy of Bacharach Instrument Company)

Figure 3-9. Boiler AM Probe Test Kit

SMOKE TESTER

To measure combustion completeness the smoke detector is used (Figure 3-7). Smoke is unburned carbon which wastes fuel, causes air pollution and fouls heat-exchanger surfaces. To use the instrument, a measured volume of flue gas is drawn through filter paper with the probe. The smoke spot is compared visually with a standard scale and a measure of smoke density is determined.

COMBUSTION ANALYZER

The combustion electronic analyzer illustrated in Figure 3-10 permits fast, close adjustments. The unit contains digital displays. A standard sampler assembly with probe (not shown) allows for stack measurements through a single stack or breaching hole.

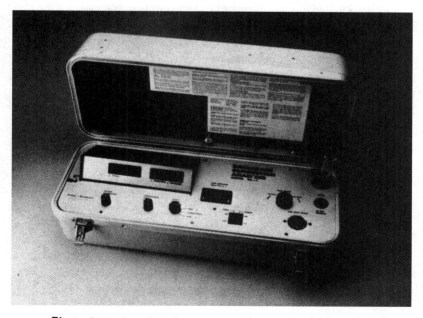

Figure 3-10. Combustion Gas Analyzer with Digital Display
(Photograph courtesy of Bacharach Instrument Company)

MEASURING HEATING, VENTILATION AND AIR-CONDITIONING (HVAC) SYSTEM PERFORMANCE

AIR VELOCITY MEASUREMENT

Table 3-2 summarizes velocity devices commonly used in HVAC applications. The following suggests the preference, suitability, and approximate costs of particular equipment.

- *Smoke pellets*—limited use but very low cost. Considered to be useful if engineering staff has experience in handling.
- *Anemometer* (deflecting vane)—good indication of air movement with acceptable order of accuracy. Considered useful. (Approx. $50).
- *Anemometer* (revolving vane)—good indicator of air movement with acceptable accuracy. However easily subject to damage. Considered useful. (Approx. $100).
- *Pitot tube*—a standard air measurement device with good levels of accuracy. Considered essential. Can be purchased in various lengths—12" about $20, 48" about $35. Must be used with a monometer. These vary considerably in cost but could be in the order of $20 to $60.
- *Impact tube*—usually packaged air flow meter kits, complete with various jets for testing ducts, grills, open areas, etc. These units are convenient to use and of sufficient accuracy. The costs vary around $150 to $300 and therefore this order of cost could only be justified for a large system.
- *Heated thermocouple*— these units are sensitive, accurate, but costly. A typical cost would be about $500 and can only be justified for regular use in a large plant.
- *Hot wire anemometer*- not recommended. Too costly and too complex.

TEMPERATURE MEASUREMENT

Table 3-3 summarizes common devices used for measuring temperature in HVAC applications. The temperature devices most commonly used are as follows:

Table 3-2. Air Velocity Measurement Devices

Device/Meter	Application	Range in FPM	Accuracy	Limitations
Smoke pellet or airborne solid tracer	Low air velocities in room —directional	5 – 50	10% – 20%	Useful in tracing air movement in-directional
Anemometer—deflecting vane type	Air velocities in rooms, grill outlets—directional	30 – 24,000	5%	Not suitable for duct air measurement—requires periodic calibration
Anemometer—revolving vane type	Moderate air velocities in ducts, rooms	100 – 3,000	5% – 20%	Subject to error variations in velocities—easily damaged. Frequent calibration required.
Pitot tube	Standard instrument for duct velocity measurement	180 – 10,000 600 – 10,000 10,000 and up	1% – 5%	Accuracy falls at low air flows.
Impact tube (side wall) meter kits	High velocity—small tube and variable direction	120 – 10,000 600 – 10,000 10,000 and up	1% – 5%	Accuracy related to constant static pressure across stream section
Heated thermocouple anemometer	Air velocities in ducts	10 – 2,000	3% – 20%	Accuracy of some meters bad at low velocities
Hot wire anemometer	(a) Low air velocities in rooms, ducts, etc. (b) High air velocity (c) Transient velocities and turbulences	1 – 1,000 Up to 60,000	1% – 20% 1% – 10%	Requires frequent calibration. Complex to use and very costly.

Table 3-3. Temperature Measurement

Device/Meter	Application	Range in °F	Accuracy °F Less than	Limitation
Glass stem thermometers	Temperature of gas, air, and liquids by contact			In gas and air, glass is affected by radiation. Also liable to break.
Mercury in glass		−38 to 575	0.1 to 10	
Alcohol in glass		−100 to 1000	0.1 to 10	
Pentane in glass		−200 to 70	0.1 to 10	
Zena or quartz mercury		−38 to 1000	0.1 to 10	
Resistance thermometers	Precision remote readings	−320 to 1800	0.02 to 5	High cost — accuracy affected by radiation
Platinum resistance	Remote readings	−150 to 300	0.03	
Nickel resistance	Remote readings	up to 600	0.1	
Thermisters				
Thermocouples				
Pt-Pt-Rh thermocouples	Standard for thermocouples	500 to 3000	0.1 to 5	Highest system
Chrome Alumel "	General testing hi-temps	up to 2000	0.1 to 15	Less accurate than above
Iron Constantan "		up to 1500	0.1 to 15	Subject to oxidation
Copper " "	Same as above but for lower readings	"	"	" " "
Chromel " "		up to 700	0.1 to 15	
Bimetallic thermometers	For approximate temperature	0 to 1000	—	Extensive time lag, not for remote use, unreliable
Pressure-bulb thermometers				Usually permanent installations. Requires careful fixing and setting
Gas filled	Suitable for remote reading	−200 to 1000	2	
Vapor filled		20 to 500	2	
Liquid filled		−50 to 2100	2	
Optical pyrometers	Hi-intensity, narrow spectrum band radiation	1500 and up	15	Limited to combustion setting
Radiation pyrometers	Hi-intensity, total high temperature radiation	Any	—	Relatively costly, easy to use, quite accurate
Indicating Crayons	Approximate surface temp.	125 to 900	+/−1%	Easy to use, low cost

- *Glass thermometers*—considered to be the most useful of temperature measuring instruments—accurate, convenient, but fragile. Cost runs from $5 each for 12" long mercury in glass. Engineers should have a selection of various ranges.
- *Resistance thermometers*—considered to be very useful for A/C testing. Accuracy is good, reliable and convenient to use. Suitable units can be purchased from $150 up, some with a selection of several temperature ranges.
- *Thermocouples*—similar to resistance thermocouple, but do not require battery power source. Chrome-Alum or iron types are the most useful and have satisfactory accuracy and repeatability. Costs start from $50 and range up.
- *Bimetallic thermometers*— considered unsuitable.
- *Pressure bulb thermometers*—more suitable for permanent installation. Accurate and reasonable in cost— $40 up.
- *Optical pyrometers*—only suitable for furnace settings and therefore limited in use. Cost from $300 up.
- *Radiation pyrometers*—limited in use for A/C work and costs from $500 up.
- *Indicating crayons*—limited in use and not considered suitable for A/C testing—costs around $2/crayon.
- *Thermographs*—use for recording room or space temperature and gives a chart indicating variations over a 12- or 168-hour period. Reasonably accurate. Low cost at around $30 to $60. (Spring wound drive.)

PRESSURE MEASUREMENT (ABSOLUTE AND DIFFERENTIAL)

Table 3-4 illustrates common devices used for measuring pressure in HVAC applications. Accuracy, range, application, and limitations are discussed in relation to HVAC work.

- *Absolute pressure manometer* ⎫ not really suited
- *Diaphragm*   ⎬   to HAVC
- *Barometer (Hg manometer)*   ⎭   test work
- *Micromanometer*—not usually portable, but suitable for

Table 3-4. Pressure Measurement

Device/Meter	Application	Range in FPM	Accuracy	Limitations
Absolute pressure manometer	Moderately low absolute pressure	0 + 30″ Hg	2 – 5%	Not direct reading
Diaphragm gauge	"	0.1 – 70 mm Hg	0.05 mm Hg	Direct reading
Barometer (Hg manometer)	Atmospheric pressure	—	0.001 to 0.01	Not very portable
Micromanometer	Very low pressure differential	0 to 6″ H_2O	0.0005 to 0.0001 H_2O	Not easily portable, hard to use with pulsating pressures
Draft gauges	Moderately low pressure differential	0 to 10″ H_2O	0.05 H_2O	Must be leveled carefully
Manometer	Medium pressure differential	0 to 100 H_2O	0.05 H_2O	Compensation for liquid density
Swing Vane gauge	Moderate low pressure differential	0 to 0.5 H_2O	5%	Generally used at atmospheric pressure only
Bourdon tube	Medium to high pressure differential. Usually to atmospheric	Any	0.05 to 5%	Subject to damage due to overpressure shock
Pressure transducers	Remote reading—responds to rapid change	0.05 to 50,000 psig	0.1 to 0.5%	Require electronic amplified and readout equipment

fixed measurement of pressure differentials across filter, coils, etc. Cost around $30 and up.

- *Draft gauges*– can be portable and used for either direct pressure or pressure differential. From $30 up.

- *Manometers*–can be portable. Used for direct pressure reading and with pitot tubes for air flows. Very useful. Costs from $20 up.

- *Swing Vane gauges*–can be portable. Usually used for air flow. Costs about $30.

- *Bourdon tube gauges*–very useful for measuring all forms of system fluid pressures from 5 psi up. Costs vary greatly from $10 up. Special types for refrigeration plants.

HUMIDITY MEASUREMENT

The data given below indicates the type of instruments available for humidity measurement. The following indicates equipment suitable for HVAC applications.

- *Psychrometers*– basically these are wet and dry bulb thermometers. They can be fixed on a portable stand or mounted in a frame with a handle for revolving in air. Costs are low ($10 to $30) and are convenient to use.

- *Dewpoint Hygrometers*–not considered suitable for HVAC test work.

- *Dimensional change*–device usually consists of a "hair" which changes in length proportionally with humidity changes. Not usually portable, fragile, and only suitable for limited temperature and humidity ranges.

- *Electrical conductivity*–can be compact and portable but of a higher cost (from $200 up). Very convenient to use.

- *Electrolytic*–as above. But for very low temperature ranges. Therefore unsuitable for HVAC test work.

- *Gravimeter* – Not suitable.

4

Energy Economic Decision Making

LIFE CYCLE COSTING

When a plant manager is assigned the role of energy manager, the first question to be asked is: "What is the economic basis for equipment purchases?"

Some companies use a simple payback method of two years or less to justify equipment purchases. Others require a life cycle cost analysis with no fuel price inflation considered. Still other companies allow for a complete life cycle cost analysis, including the impact for the fuel price inflation and the energy tax credit.

The energy manager's success is directly related to how he or she must justify energy utilization methods.

USING THE PAYBACK PERIOD METHOD

The payback period is the time required to recover the capital investment out of the earnings or savings. This method ignores all savings beyond the payback years, thus penalizing projects that have long life potentials for those that offer high savings for a relatively short period.

The payback period criterion is used when funds are limited and it is important to know how fast dollars will come back. The payback period is simply computed as:

$$\text{Payback period} = \frac{\text{initial investment}}{\text{after tax savings}}$$

(4-1)

The energy manager who must justify energy equipment expenditures based on a payback period of one year or less has little chance for long-range success. Some companies have set higher payback periods for energy utilization methods. These longer payback periods are justified on the basis that:

- Fuel pricing will increase at a higher rate than the general inflation rate.
- The "risk analysis" for not implementing energy utilization measures may mean loss of production and losing a competitive edge.

USING LIFE CYCLE COSTING

Life cycle costing is an analysis of the total cost of a system, device, building, machine, etc., over its anticipated useful life. The name is new but the subject has, in the past, gone by such names as "engineering economic analysis" or "total owning and operating cost summaries."

Life cycle costing has brought about a new emphasis on the comprehensive identification of all costs associated with a system. The most commonly included costs are initial in place cost, operating costs, maintenance costs, and interest on the investment. Two factors enter into appraising the life of the system; namely, the expected physical life and the period of obsolescence. The lesser factor is governing time period. The effect of interest can then be calculated by using one of several formulas which take into account the time value of money.

When comparing alternative solutions to a particular problem, the system showing the lowest life cycle cost will usually be the first choice (performance requirements are assessed as equal in value).

Life cycle costing is a tool in value engineering. Other items, such as installation time, pollution effects, aesthetic considerations, delivery time, and owner preferences will temper the rule of always choosing the system with the lowest life cycle cost. Good overall judgement is still required.

The life cycle cost analysis still contains judgement factors pertaining to interest rates, useful life, and inflation rates. Even with the judgement element, life cycle costing is the most important tool in value engineering, since the results are quantified in terms of dollars.

As the price for energy changes, and as governmental incentives are initiated, processes or alternatives which were not economically feasible will be considered. This chapter will concentrate on the principles of the life cycle cost analysis as they apply to energy conservation decision making.

THE TIME VALUE OF MONEY

Most energy saving proposals require the investment of capital to accomplish them. By investing today in energy conservation, yearly operating dollars over the life of the investment will be saved. A dollar in hand today is more valuable than one to be received at some time in the future. For this reason, a *time value* must be placed on all cash flows into and out of the company.

Money transactions are thought of as a cash flow to or from a company. Investment decisions also take into account alternate investment opportunities and the minimum return on the investment. In order to compute the rate of return on an investment, it is necessary to find the interest rate which equates payments outcoming and incoming, present and future. The method used to find the rate of return is referred to as *discounted cashflow*.

INVESTMENT DECISION-MAKING

To make investment decisions, the energy manager must follow one simple principle: Relate annual cash flows and lump sum deposits to the same time base. The six factors used for investment decision making simply convert cash from one time base to another; since each company has various financial objectives, these factors can be used to solve *any* investment problem.

Single Payment Compound Amount—F/P

Use the F/P factor to determine future amount F that present sum P will accumulate at i percent interest, in n years. If P (present worth) is known, and F (future worth) is to be determined, use Equation 4-2.

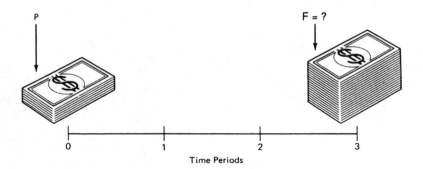

Fig. 4-1. Single payment compound amount (F/P).

$$F = P \times (1 + i)^n \qquad (4\text{-}2)$$

$$F/P = (1 + i)^n \qquad (4\text{-}3)$$

The F/P can be computed by an interest formula, but usually its value is found by using the interest tables. Interest tables for interest rates of 10 to 50 percent are found at the conclusion of this chapter (Tables 4-1 through 4-8). In predicting future costs, there are many unknowns. For the accuracy of most calculations, interest rates are assumed to be compounded annually unless otherwise specified. Linear interpolation is commonly used to find values not listed in the interest tables.

Tables 4-9 through 4-12 can be used to determine the effect of fuel escalation on the life cycle cost analysis.

Single Payment Present Worth—P/F

The P/F factor is used to determine the present worth, P, that a future amount, F, will be at interest of i-percent, in n years. If F is known, and P is to be determined, then Equation 4-4 is used.

$$P = F \times 1/(1 + i)^n \qquad (4\text{-}4)$$

$$P/F = \frac{1}{(1 + i)^n} \qquad (4\text{-}5)$$

TABLE 4-1. 10% Interest Factors.

Period n	Single-payment compound-amount F/P	Single-payment present-worth P/F	Uniform-series compound-amount F/A	Sinking-fund payment A/F	Capital recovery A/P	Uniform-series present-worth P/A
	Future value of $1 $(1 + i)^n$	Present value of $1 $\dfrac{1}{(1 + i)^n}$	Future value of uniform series of $1 $\dfrac{(1 + i)^n - 1}{i}$	Uniform series whose future value is $1 $\dfrac{i}{(1 + i)^n - 1}$	Uniform series with present value of $1 $\dfrac{i(1+i)^n}{(1 + i)^n - 1}$	Present value of uniform series of $1 $\dfrac{(1+i)^n - 1}{i(1 + i)^n}$
1	1.100	0.9091	1.000	1.00000	1.10000	0.909
2	1.210	0.8264	2.100	0.47619	0.57619	1.736
3	1.331	0.7513	3.310	0.30211	0.40211	2.487
4	1.464	0.6830	4.641	0 21547	0.31147	3.170
5	1.611	0.6209	6.105	0.16380	0.26380	3.791
6	1.772	0.5645	7.716	0.12961	0.22961	4.355
7	1.949	0.5132	9.487	0.10541	0.20541	4.868
8	2.144	0.4665	11.436	0.08744	0.18744	5.335
9	2.358	0.4241	13.579	0.07364	0.17364	5.759
10	2.594	0.3855	15.937	0.06275	0.16275	6.144
11	2.853	0.3505	18.531	0.05396	0.15396	6.495
12	3.138	0.3186	21.384	0.04676	0.14676	6.814
13	3.452	0.2897	24.523	0.04078	0.14078	7 103
14	3.797	0.2633	27.975	0.03575	0.13575	7 367
15	4.177	0.2394	31.772	0.03147	0.13147	7 f)06
16	4.595	0 2176	35.950	0.02782	0.12782	7.824
17	5.054	0.1978	40.545	0.02466	0.12466	8.022
18	5.560	0.1799	45.599	0.02193	0.12193	8.201
19	6.116	0.1635	51.159	0.01955	0.11955	8.365
20	6.727	0.1486	57.275	0.01746	0.11746	8.514
21	7.400	0.1351	64.002	0.01562	0.11562	8.649
22	8.140	0.1228	71.403	0.01401	0.11401	8.772
23	8.954	0.1117	79.543	0.01257	0.11257	8 883
24	9.850	0.1015	88.497	0.01130	0.11130	8.985
25	10.835	0.0923	98.347	0.01017	0.11017	9.077
26	11.918	0.0839	109.182	0.00916	0 10916	9.161
27	13.110	0.0763	121.100	0.00826	0.10826	9.237
28	14.421	0.0693	134.210	0.00745	0.10745	9.307
29	15.863	0.0630	148.631	0.00673	0.10673	9.370
30	17.449	0.0673	164.494	0.00608	0.10608	9.427
35	28.102	0.0356	271.024	0.00369	0.10369	9.644
40	45.259	0.0221	442.593	0.00226	0.10226	9.779
45	72.890	0.0137	718.905	0.00139	0.10139	9.863
50	117.391	0.0085	1163.909	0.00086	0.10086	9.915
55	189.059	0.0053	1880.591	0.00053	0.10053	9.947
60	304.482	0.0033	3034.816	0.00033	0.10033	9.967
65	490.371	0.0020	4893.902	0.00020	0.10020	9.980
70	789.747	0.0013	7887.470	0.00013	0.10013	9.987
75	1271.895	0.0008	12708.954	0.00008	0.10008	9.992
80	2048.400	0.0005	20474.002	0.00005	0.10005	9.995
85	3298. 969	0.0003	32979.690	0.00003	0 10003	9.997
90	5313.023	0.0002	53120.226	0.00002	0.10002	9.998
95	8556.676	0.0001	85556.760	0.00001	0.10001	9.999

TABLE 4-2. 12% Interest Factors.

Period n	Single-payment compound-amount F/P	Single-payment present-worth P/F	Uniform-series compound-amount F/A	Sinking-fund payment A/F	Capital recovery A/P	Uniform-series present-worth P/A
	Future value of $1 $(1 + i)^n$	Present value of $1 $\dfrac{1}{(1+i)^n}$	Future value of uniform series of $1 $\dfrac{(1+i)^n-1}{i}$	Uniform series whose future value is $1 $\dfrac{i}{(1+i)^n-1}$	Uniform series with present value of $1 $\dfrac{i(1+i)^n}{(1+i)^n-1}$	Present value of uniform series of $1 $\dfrac{(1+i)^n-1}{i(1+i)^n}$
1	1.120	0.8929	1.000	1.00000	1.12000	0.893
2	1.254	0.7972	2.120	0.47170	0.59170	1.690
3	1.405	0.7118	3.374	0.29635	0.41635	2.402
4	1.574	0.6355	4.779	0.20923	0.32923	3.037
5	1.762	0.5674	6.353	0.15741	0.27741	3.605
6	1.974	0.5066	8.115	0.12323	0.24323	4.111
7	2.211	0.4523	10.089	0.09912	0.21912	4.564
8	2.476	0.4039	12.300	0.08130	0.20130	4.968
9	2.773	0.3606	14.776	0.06768	0.18768	5.328
10	3.106	0.3220	17.549	0.05698	0.17698	5.650
11	3.479	0.2875	20.655	0.04842	0.16842	5.938
12	3.896	0.2567	24.133	0.04144	0.16144	6.194
13	4.363	0.2292	28.029	0.03568	0.15568	6.424
14	4.887	0.2046	32.393	0.03087	0.15087	6.628
15	5.474	0.1827	37.280	0.02682	0.14682	6.811
16	6.130	0.1631	42.753	0.02339	0.14339	6.974
17	6.866	0.1456	48.884	0.02046	0.14046	7.120
18	7.690	0.1300	55.750	0.01794	0.13794	7.250
19	8.613	0.1161	63.440	0.01576	0.13576	7.366
20	9.646	0.1037	72.052	0.01388	0.13388	7.469
21	10.804	0.0926	81.699	0.01224	0.13224	7.562
22	12.100	0.0826	92.503	0.01081	0.13081	7.645
23	13.552	0.0738	104.603	0.00956	0.12956	7.718
24	15.179	0.0659	118.155	0.00846	0.12846	7.784
25	17.000	0.0588	133.334	0.00750	0.12750	7.843
26	19.040	0.0525	150.334	0.00665	0.12665	7.896
27	21.325	0.0469	169.374	0.00590	0.12590	7.943
28	23.884	0.0419	190.699	0.00524	0.12524	7.984
29	26.750	0.0374	214.583	0.00466	0.12466	8.022
30	29.960	0.0334	241.333	0.00414	0.12414	8.055
35	52.800	0.0189	431.663	0.00232	0.12232	8.176
40	93.051	0.0107	767.091	0.00130	0.12130	8.244
45	163.988	0.0061	1358.230	0.00074	0.12074	8.283
50	289.002	0.0035	2400.018	0.00042	0.12042	8.304
55	509.321	0.0020	4236.005	0.00024	0.12024	8.317
60	897.597	0.0011	7471.641	0.00013	0.12013	8.324
65	1581.872	0.0006	13173.937	0.00008	0.12008	8.328
70	2787.800	0.0004	23223.332	0.00004	0.12004	8.330
75	4913.056	0.0002	40933.799	0.00002	0.12002	8.332
80	8658.483	0.0001	72145.692	0.00001	0.12001	8.332

TABLE 4-3. 15% Interest Factors.

Period n	Single-payment compound-amount F/P	Single-payment present-worth P/F	Uniform-series compound-amount F/A	Sinking-fund payment A/F	Capital recovery A/P	Uniform-series present-worth P/A
	Future value of $1 $(1+i)^n$	Present value of $1 $\dfrac{1}{(1+i)^n}$	Future value of uniform series of $1 $\dfrac{(1+i)^n-1}{i}$	Uniform series whose future value is $1 $\dfrac{i}{(1+i)^n-1}$	Uniform series with present value of $1 $\dfrac{i(1+i)^n}{(1+i)^n-1}$	Present value of uniform series of $1 $\dfrac{(1+i)^n-1}{i(1+i)^n}$
1	1.150	0.8696	1.000	1.00000	1.15000	0.870
2	1.322	0.7561	2.150	0.46512	0.61512	1.626
3	1.521	0.6575	3.472	0.28798	0.43798	2.283
4	1.749	0.5718	4.993	0.20027	0.35027	2.855
5	2.011	0.4972	6.742	0.14832	0.29832	3.352
6	2.313	0.4323	8.754	0.11424	0.26424	3.784
7	2.660	0.3759	11.067	0.09036	0.24036	4.160
8	3.059	0.3269	13.727	0.07285	0.22285	4.487
9	3.518	0.2843	16.786	0.05957	0.20957	4.772
10	4.046	0.2472	20.304	0.04925	0.19925	5.019
11	4.652	0.2149	24.349	0.04107	0.19107	5.234
12	5.350	0.1869	29.002	0.03448	0.18448	5.421
13	6.153	0.1625	34.352	0.02911	0.17911	5.583
14	7.076	0.1413	40.505	0.02469	0.17469	5.724
15	8.137	0.1229	47.580	0.02102	0.17102	5.847
16	9.358	0.1069	55.717	0.01795	0.16795	5.954
17	10.761	0.0929	65.075	0.01537	0.16537	6.047
18	12.375	0.0808	75.836	0.01319	0.16319	6.128
19	14.232	0.0703	88.212	0.01134	0.16134	6.198
20	16.367	0.0611	102.444	0.00976	0.15976	6.259
21	18.822	0.0531	118.810	0.00842	0.15842	6.312
22	21.645	0.0462	137.632	0.00727	0.15727	6.359
23	24.891	0.0402	159.276	0.00628	0.15628	6.399
24	28.625	0.0349	184.168	0.00543	0.15543	6.434
25	32.919	0 0304	212.793	0.00470	0.15470	6.464
26	37.857	0.0264	245 712	0.00407	0.15407	6.491
27	43.535	0.0230	283.569	0.00353	0.15353	6.514
28	50.066	0 0200	327.104	0.00306	0.15306	6.534
29	57.575	0.0174	377.170	0.00265	0.15265	6.551
30	66.212	0.0151	434.745	0.00230	0.15230	6.566
35	133.176	0.0075	881.170	0.00113	0.15113	6.617
40	267.864	0.0037	1779.090	0.00056	0.15056	6.642
45	538.769	0.0019	3585.128	0.00028	0.15028	6.654
50	1083.657	0.0009	7217.716	0.00014	0.15014	6.661
55	2179.622	0.0005	14524.148	0.00007	0.15007	6.664
60	4383.999	0.0002	29219.992	0.00003	0.15003	6.665
65	8817.787	0.0001	58778.583	0.00002	0.15002	6.666

TABLE 4-4. 20% Interest Factors.

Period n	Single-payment compound-amount F/P	Single-payment present-worth P/F	Uniform-series compound-amount F/A	Sinking-fund payment A/F	Capital recovery A/P	Uniform-series present-worth P/A
	Future value of $1 $(1+i)^n$	Present value of $1 $\dfrac{1}{(1+i)^n}$	Future value of uniform series of $1 $\dfrac{(1+i)^n-1}{i}$	Uniform series whose future value is $1 $\dfrac{i}{(1+i)^n-1}$	Uniform series with present value of $1 $\dfrac{i(1+i)^n}{(1+i)^n-1}$	Present value of uniform series of $1 $\dfrac{(1+i)^n-1}{i(1+i)^n}$
1	1.200	0.8333	1.000	1.00000	1.20000	0.833
2	1.440	0.6944	2.200	0.45455	0.65455	1.528
3	1.728	0.5787	3.640	0.27473	0.47473	2.106
4	2.074	0.4823	5.368	0.18629	0.38629	2.589
5	2.488	0.4019	7.442	0.13438	0.33438	2.991
6	2.986	0.3349	9.930	0.10071	0.30071	3.326
7	3.583	0.2791	12.916	0.07742	0.27742	3.605
8	4 300	0.2326	16.499	0.06061	0.26061	3.837
9	5.160	0.1938	20.799	0.04808	0.24808	4.031
10	6 192	0.1615	25.959	0.03852	0.23852	4.192
11	7.430	0.1346	32.150	0.03110	0.23110	4.327
12	8.916	0.1122	39.581	0.02526	0.22526	4.439
13	10.699	0.0935	48.497	0.02062	0.22062	4.533
14	12.839	0.0779	59.196	0.01689	0.21689	4.611
15	15.407	0.0649	72.035	0.01388	0.21388	4.675
16	18.488	0.0541	87.442	0.01144	0.21144	4.730
17	22.186	0.0451	105.931	0.00944	0.20944	4.775
18	26.623	0.0376	128.117	0.00781	0.20781	4.812
19	31.948	0.0313	154.740	0.00646	0.20646	4.843
20	38.338	0.0261	186.688	0.00536	0.20536	4.870
21	46.005	0.0217	225.026	0 00444	0.20444	4.891
22	55.206	0.0181	271.031	0.00369	0.20369	4.909
23	66.247	0.0151	326.237	0.00307	0.20307	4.925
24	79.497	0.0126	392.484	0.00255	0.20255	4.937
25	95.396	0.0105	471.981	0.00212	0.20212	4.948
26	114.475	0.0087	567.377	0.00176	0.20176	4.956
27	137.371	0.0073	681.853	0.00147	0.20147	4.964
28	164.845	0.0061	819.223	0.00122	0.20122	4.970
29	197.814	0.0051	984.068	0.00102	0.20102	4.975
30	237.376	0.0042	1181.882	0.00085	0.20085	4.979
35	590.668	0.0017	2948.341	0.00034	0.20034	4.992
40	1469.772	0.0007	7343.858	0.00014	0.20014	4.997
45	3657.262	0.0003	18281 310	0.00005	0.20005	4.999
50	9100.438	0.0001	45497.191	0.00002	0 20002	4 999

TABLE 4-5. 25% Interest Factors.

Period n	Single-payment compound-amount F/P	Single-payment present-worth P/F	Uniform-series compound-amount F/A	Sinking-fund payment A/F	Capital recovery A/P	Uniform-series present-worth P/A
	Future value of $1 $(1+i)^n$	Present value of $1 $\dfrac{1}{(1+i)^n}$	Future value of uniform series of $1 $\dfrac{(1+i)^n-1}{i}$	Uniform series whose future value is $1 $\dfrac{i}{(1+i)^n-1}$	Uniform series with present value of $1 $\dfrac{i(1+i)^n}{(1+i)^n-1}$	Present value of uniform series of $1 $\dfrac{(1+i)^n-1}{i(1+i)^n}$
1	1.250	0.8000	1.000	1.00000	1.25000	0.800
2	1.562	0.6400	2.250	0.44444	0.69444	1.440
3	1.953	0.5120	3.812	0.26230	0.51230	1.952
4	2.441	0.4096	5.766	0.17344	0.42344	2.362
5	3.052	0.3277	8.207	0.12185	0.37185	2.689
6	3.815	0.2621	11.259	0.08882	0.33882	2.951
7	4.768	0.2097	15.073	0.06634	0.31634	3.161
8	5.960	0.1678	19.842	0.05040	0.30040	3.329
9	7.451	0 1342	25.802	0.03876	0.28876	3.463
10	9.313	0.1074	33.253	0.03007	0.28007	3.571
11	11.642	0.0859	42.566	0.02349	0.27349	3.656
12	14.552	0.0687	54.208	0.01845	0.26845	3.725
13	18.190	0.0550	68.760	0.01454	0.26454	3.780
14	22.737	0.0440	86.949	0.01150	0.26150	3.824
15	28.422	0.0352	109.687	0.00912	0.25912	3.859
16	35.527	0.0281	138.109	0.00724	0.25724	3.887
17	44.409	0.0225	173.636	0.00576	0.25576	3.910
18	55.511	0.0180	218.045	0.00459	0.25459	3.928
19	69.389	0.0144	273.556	0.00366	0.25366	3.942
20	86.736	0.0115	342.945	0.00292	0.25292	3.954
21	108.420	0.0092	429 681	0.00233	0.25233	3.963
22	135.525	0.0074	538.101	0.00186	0.25186	3.970
23	169.407	0.0059	673.626	0.00148	0.25148	3.976
24	211.758	0.0047	843.033	0.00119	0.25119	3.981
25	264.698	0.0038	1054.791	0.00095	0.25095	3.985
26	330.872	0.0030	1319.489	0.00076	0.25076	3.988
27	413.590	0.0024	1650.361	0.00061	0.25061	3.990
28	516.988	0.0019	2063.952	0.00048	0.25048	3.992
29	646.235	0.0015	2580.939	0.00039	0.25039	3.994
30	807.794	0.0012	3227. 174	0.00031	0.25031	3.995
35	2465.190	0.0004	9856.761	0.00010	0.25010	3.998
40	7523.164	0.0001	30088.655	0.00003	0.25003	3.999

TABLE 4-6. 30% Interest Factors.

Period n	Single-payment compound-amount F/P Future value of $1 $(1+i)^n$	Single-payment present-worth P/F Present value of $1 $\dfrac{1}{(1+i)^n}$	Uniform-series compound-amount F/A Future value of uniform series of $1 $\dfrac{(1+i)^n-1}{i}$	Sinking-fund payment A/F Uniform series whose future value is $1 $\dfrac{i}{(1+i)^n-1}$	Capital recovery A/P Uniform series with present value of $1 $\dfrac{i(1+i)^n}{(1+i)^n-1}$	Uniform-series present-worth P/A Present value of uniform series of $1 $\dfrac{(1+i)^n-1}{i(1+i)^n}$
1	1.300	0.7692	1.000	1.00000	1.30000	0.769
2	1.690	0.5917	2.300	0.43478	0.73478	1.361
3	2.197	0.4552	3.990	0.25063	0.55063	1.816
4	2.856	0.3501	6.187	0.16163	0.46163	2.166
5	3.713	0.2693	9.043	0.11058	0.41058	2.436
6	4.827	0.2072	12.756	0.07839	0.37839	2.643
7	6.275	0.1594	17.583	0.05687	0.35687	2.802
8	8.157	0.1226	23.858	0.04192	0.34192	2.925
9	10.604	0.0943	32.015	0.03124	0.33124	3.019
10	13.786	0.0725	42.619	0.02346	0.32346	3.092
11	17.922	0.0558	56.405	0.01773	0.31773	3.147
12	23.298	0.0429	74.327	0.01345	0.31345	3.190
13	30.288	0.0330	97.625	0.01024	0.31024	3.223
14	39.374	0.0254	127.913	0.00782	0.30782	3.249
15	51.186	0.0195	167.286	0.00598	0.30598	3.268
16	66.542	0.0150	218.472	0.00458	0.30458	3.283
17	86.504	0.0116	285.014	0.00351	0.30351	3.295
18	112.455	0.0089	371.518	0.00269	0.30269	3.304
19	146.192	0.0068	483.973	0.00207	0.30207	3.311
20	190.050	0.0053	630.165	0.00159	0.30159	3.316
21	247.065	0.0040	820 215	0.00122	0.30122	3.320
22	321.184	0.0031	1067.280	0.00094	0.30094	3.323
23	417.539	0.0024	1388.464	0.00072	0.30072	3.325
24	542.801	0.0018	1806.003	0.00055	0.30055	3.327
25	705.641	0.0014	2348.803	0.00043	0 30043	3.329
26	917.333	0.0011	3054.444	0.00033	0.30033	3.330
27	1192.533	0.0008	3971.778	0.00025	0.30025	3.331
28	1550.293	0.0006	5164.311	0.00019	0.30019	3.331
29	2015.381	0.0005	6714.604	0.00015	0.30015	3.332
30	2619.996	0.0004	8729.985	0.00011	0.30011	3.332
35	9727.8060	0.0001	32422.868	0.00003	0.30003	3.333

TABLE 4-7. 40% Interest Factors.

Period n	Single-payment compound-amount F/P	Single-payment present-worth P/F	Uniform-series compound-amount F/A	Sinking-fund payment A/F	Capital recovery A/P	Uniform-series present-worth P/A
	Future value of $1 $(1 + i)^n$	Present value of $1 $\dfrac{1}{(1 + i)^n}$	Future value of uniform series of $1 $\dfrac{(1 + i)^n - 1}{i}$	Uniform series whose future value is $1 $\dfrac{i}{(1 + i)^n - 1}$	Uniform series with present value of $1 $\dfrac{i(1+i)^n}{(1 + i)^n - 1}$	Present value of uniform series of $1 $\dfrac{(1+i)^n - 1}{i(1 + i)^n}$
1	1.400	0.7143	1.000	1.00000	1.40000	0.714
2	1.960	0.5102	2.400	0.41667	0.81667	1.224
3	2.744	0.3644	4.360	0 22936	0.62936	1.589
4	3.842	0.2603	7.104	0.14077	0.54077	1.849
5	5.378	0.1859	10.946	0.09136	0.49136	2.035
6	7.530	0.1328	16.324	0.06126	0.46126	2.168
7	10.541	0.0949	23.853	0.04192	0.44192	2.263
8	14.758	0.0678	34.395	0.02907	0.42907	2.331
9	20.661	0.0484	49.153	0.02034	0.42034	2.379
10	28.925	0.0346	69.814	0.01432	0.41432	2.414
11	40.496	0.0247	98.739	0.01013	0.41013	2.438
12	56.694	0.0176	139.235	0.00718	0.40718	2.456
13	79.371	0.0126	195.929	0.00510	0.40510	2.469
14	111.120	0.0090	275.300	0.00363	0.40363	2.478
15	155.568	0.0064	386.420	0.00259	0.40259	2.484
16	217.795	0.0046	541.988	0.00185	0.40185	2.489
17	304.913	0.0033	759.784	0.00132	0.40132	2.492
18	426.879	0.0023	1064.697	0.00094	0.40094	2.494
19	597.630	0.0017	1491.576	0.00067	0.40067	2.496
20	836.683	0.0012	2089.206	0.00048	0.40048	2.497
21	1171.356	0.0009	2925.889	0.00034	0.40034	2.498
22	1639.898	0.0006	4097.245	0.00024	0.40024	2.498
23	2295.857	0.0004	5737.142	0.00017	0.40017	2.499
24	3214.200	0.0003	8032.999	0.00012	0.40012	2.499
25	4499.880	0.0002	11247.199	0.00009	0.40009	2.499
26	6299.831	0.0002	15747.079	0.00006	0.40006	2.500
27	8819.764	0.0001	22046.910	0.00005	0.40005	2.500

TABLE 4-8. 50% Interest Factors.

Period n	Single-payment compound-amount F/P	Single-payment present-worth P/F	Uniform-series compound-amount F/A	Sinking-fund payment A/F	Capital recovery A/P	Uniform-series present-worth P/A
	Future value of $1 $(1+i)^n$	Present value of $1 $\dfrac{1}{(1+i)^n}$	Future value of uniform series of $1 $\dfrac{(1+i)^n-1}{i}$	Uniform series whose future value is $1 $\dfrac{i}{(1+i)^n-1}$	Uniform series with present value of $1 $\dfrac{i(1+i)^n}{(1+i)^n-1}$	Present value of uniform series of $1 $\dfrac{(1+i)^n-1}{i(1+i)^n}$
1	1.500	0.6667	1.000	1.00000	1.50000	0.667
2	2.250	0.4444	2.500	0.40000	0.90000	1.111
3	3.375	0.2963	4.750	0.21053	0.71053	1.407
4	5.062	0.1975	8.125	0.12308	0.62308	1.605
5	7.594	0.1317	13.188	0.07583	0.57583	1.737
6	11.391	0.0878	20.781	0.04812	0.54812	1.824
7	17.086	0.0585	32.172	0.03108	0.53108	1.883
8	25.629	0.0390	49.258	0.02030	0.52030	1.922
9	38.443	0.0260	74.887	0.01335	0.51335	1.948
10	57.665	0.0173	113.330	0.00882	0.50882	1.965
11	86.498	0.0116	170.995	0.00585	0.50585	1.977
12	129.746	0.0077	257.493	0.00388	0.50388	1.985
13	194.620	0.0051	387.239	0.00258	0.50258	1.990
14	291.929	0.0034	581.859	0.00172	0.50172	1.993
15	437.894	0.0023	873.788	0.00114	0.50114	1.995
16	656.841	0.0015	1311.682	0.00076	0.50076	1.997
17	985.261	0.0010	1968.523	0.00051	0.50051	1.998
18	1477.892	0.0007	2953.784	0.00034	0.50034	1.999
19	2216.838	0.0005	4431.676	0.00023	0.50023	1.999
20	3325.257	0.0003	6648.513	0.00015	0.50015	1.999
21	4987.885	0.0002	9973.770	0.00010	0.50010	2.000
22	7481.828	0.0001	14961.655	0.00007	0.50007	2.000

TABLE 4-9. Five-Year Escalation Table.

Present Worth of a Series of Escalating Payments Compounded Annually
Discount-Escalation Factors for $n = 5$ Years

Discount Rate	Annual Escalation Rate					
	0.10	0.12	0.14	0.16	0.18	0.20
0.10	5.000000	5.279234	5.572605	5.880105	6.202627	6.540569
0.11	4.866862	5.136200	5.420152	5.717603	6.029313	6.355882
0.12	4.738562	5.000000	5.274242	5.561868	5.863289	6.179066
0.13	4.615647	4.869164	5.133876	5.412404	5.704137	6.009541
0.14	4.497670	4.742953	5.000000	5.269208	5.551563	5.847029
0.15	4.384494	4.622149	4.871228	5.131703	5.404955	5.691165
0.16	4.275647	4.505953	4.747390	5.000000	5.264441	5.541511
0.17	4.171042	4.394428	4.628438	4.873699	5.129353	5.397964
0.18	4.070432	4.287089	4.513947	4.751566	5.000000	5.259749
0.19	3.973684	4.183921	4.403996	4.634350	4.875619	5.126925
0.20	3.880510	4.084577	4.298207	4.521178	4.755725	5.000000
0.21	3.790801	3.989001	4.196400	4.413341	4.640260	4.877689
0.22	4.704368	3.896891	4.098287	4.308947	4.529298	4.759649
0.23	3.621094	3.808179	4.003835	4.208479	4.422339	4.645864
0.24	3.540773	3.722628	3.912807	4.111612	4.319417	4.536517
0.25	3.463301	3.640161	3.825008	4.018249	4.220158	4.431144
0.26	3.388553	3.560586	3.740376	3.928286	4.124553	4.329514
0.27	3.316408	3.483803	3.658706	3.841442	4.032275	4.231583
0.28	3.246718	3.409649	3.579870	3.757639	3.943295	4.137057
0.29	3.179393	3.338051	3.503722	3.676771	3.857370	4.045902
0.30	3.114338	3.268861	3.430201	3.598653	3.774459	3.957921
0.31	3.051452	3.201978	3.359143	3.523171	3.694328	3.872901
0.32	2.990618	3.137327	3.290436	3.450224	3.616936	3.790808
0.33	2.931764	3.074780	3.224015	3.379722	3.542100	3.711472
0.34	2.874812	3.014281	3.159770	3.311524	3.469775	3.634758

TABLE 4-10. Ten-Year Escalation Table.

Present Worth of a Series of Escalating Payments Compounded Annually
Discount-Escalation Factors for $n = 10$ Years

Discount Rate	Annual Escalation Rate					
	0.10	0.12	0.14	0.16	0.18	0.20
0.10	10.000000	11.056250	12.234870	13.548650	15.013550	16.646080
0.11	9.518405	10.508020	11.613440	12.844310	14.215140	15.741560
0.12	9.068870	10.000000	11.036530	12.190470	13.474590	14.903510
0.13	8.650280	9.526666	10.498990	11.582430	12.786980	14.125780
0.14	8.259741	9.084209	10.000000	11.017130	12.147890	13.403480
0.15	7.895187	8.672058	9.534301	10.490510	11.552670	12.731900
0.16	7.554141	8.286779	9.099380	10.000000	10.998720	12.106600
0.17	7.234974	7.926784	8.693151	9.542653	10.481740	11.524400
0.18	6.935890	7.589595	8.312960	9.113885	10.000000	10.980620
0.19	6.655455	7.273785	7.957330	8.713262	9.549790	10.472990
0.20	6.392080	6.977461	7.624072	8.338518	9.128122	10.000000
0.21	6.144593	6.699373	7.311519	7.987156	8.733109	9.557141
0.22	5.911755	6.437922	7.017915	7.657542	8.363208	9.141752
0.23	5.692557	6.192047	6.742093	7.348193	8.015993	8.752133
0.24	5.485921	5.960481	6.482632	7.057347	7.690163	8.387045
0.25	5.290990	5.742294	6.238276	6.783767	7.383800	8.044173
0.26	5.106956	5.536463	6.008083	6.526298	7.095769	7.721807
0.27	4.933045	5.342146	5.790929	6.283557	6.824442	7.418647
0.28	4.768518	5.158489	5.585917	6.054608	6.568835	7.133100
0.29	4.612762	4.984826	5.392166	5.838531	6.327682	6.864109
0.30	4.465205	4.820429	5.209000	5.634354	6.100129	6.610435
0.31	4.325286	4.664669	5.035615	5.441257	5.885058	6.370867
0.32	4.192478	4.517015	4.871346	5.258512	5.681746	6.144601
0.33	4.066339	4.376884	4.715648	5.085461	5.489304	5.930659
0.34	3.946452	4.243845	4.567942	4.921409	5.307107	5.728189

TABLE 4-11. Fifteen-Year Escalation Table.

Present Worth of a Series of Escalating Payments Compounded Annually
Discount-Escalation Factors for n = 15 years

Discount Rate	Annual Escalation Rate					
	0.10	0.12	0.14	0.16	0.18	0.20
0.10	15.000000	17.377880	20.199780	23.549540	27.529640	32.259620
0.11	13.964150	16.126230	18.690120	21.727370	25.328490	29.601330
0.12	13.026090	15.000000	17.332040	20.090360	23.355070	27.221890
0.13	12.177030	13.981710	16.105770	18.616160	21.581750	25.087260
0.14	11.406510	13.057790	15.000000	17.287320	19.985530	23.169060
0.15	10.706220	12.220570	13.998120	16.086500	18.545150	21.442230
0.16	10.068030	11.459170	13.088900	15.000000	17.244580	19.884420
0.17	9.485654	10.766180	12.262790	14.015480	16.066830	18.477610
0.18	8.953083	10.133630	11.510270	13.118840	15.000000	17.203010
0.19	8.465335	9.555676	10.824310	12.303300	14.030830	16.047480
0.20	8.017635	9.026333	10.197550	11.560150	13.148090	15.000000
0.21	7.606115	8.540965	9.623969	10.881130	12.343120	14.046400
0.22	7.227109	8.094845	9.097863	10.259820	11.608480	13.176250
0.23	6.877548	7.684317	8.614813	9.690559	10.936240	12.381480
0.24	6.554501	7.305762	8.170423	9.167798	10.320590	11.655310
0.25	6.255518	6.956243	7.760848	8.687104	9.755424	10.990130
0.26	5.978393	6.632936	7.382943	8.244519	9.236152	10.379760
0.27	5.721101	6.333429	7.033547	7.836080	8.757889	9.819020
0.28	5.481814	6.055485	6.710042	7.458700	8.316982	9.302823
0.29	5.258970	5.797236	6.410005	7.109541	7.909701	8.827153
0.30	5.051153	5.556882	6.131433	6.785917	7.533113	8.388091
0.31	4.857052	5.332839	5.872303	6.485500	7.184156	7.982019
0.32	4.675478	5.123753	5.630905	6.206250	6.860492	7.606122
0.33	4.505413	4.928297	5.405771	5.946343	6.559743	7.257569
0.34	4.345926	4.745399	5.195502	5.704048	6.280019	6.933897

TABLE 4-12. Twenty-Year Escalation Table.

Present Worth of a Series of Escalating Payments Compounded Annually
Discount-Escalation Factors for n = 20 Years

Discount Rate	Annual Escalation Rate					
	0.10	0.12	0.14	0.16	0.18	0.20
0.10	20.000000	24.295450	29.722090	36.592170	45.308970	56.383330
0.11	18.213210	22.002090	26.776150	32.799710	40.417480	50.067940
0.12	16.642370	20.000000	24.210030	29.505400	36.181240	44.614710
0.13	15.259850	18.243100	21.964990	26.634490	32.502270	39.891400
0.14	14.038630	16.694830	20.000000	24.127100	29.298170	35.789680
0.15	12.957040	15.329770	18.271200	21.929940	26.498510	32.218060
0.16	11.995640	14.121040	16.746150	20.000000	24.047720	29.098950
0.17	11.138940	13.048560	15.397670	18.300390	21.894660	26.369210
0.18	10.373120	12.093400	14.201180	16.795710	20.000000	23.970940
0.19	9.686791	11.240870	13.137510	15.463070	18.326720	21.860120
0.20	9.069737	10.477430	12.188860	14.279470	16.844020	20.000000
0.21	8.513605	9.792256	11.340570	13.224610	15.527270	18.353210
0.22	8.010912	9.175267	10.579620	12.282120	14.355520	16.890730
0.23	7.555427	8.618459	9.895583	11.438060	13.309280	15.589300
0.24	7.141531	8.114476	9.278916	10.679810	12.373300	14.429370
0.25	6.764528	7.657278	8.721467	9.997057	11.533310	13.392180
0.26	6.420316	7.241402	8.216490	9.380883	10.778020	12.462340
0.27	6.105252	6.862203	7.757722	8.823063	10.096710	11.626890
0.28	5.816151	6.515563	7.339966	8.316995	9.480940	10.874120
0.29	5.550301	6.198027	6.958601	7.856833	8.922847	10.194520
0.30	5.305312	5.906440	6.609778	7.437339	8.416060	9.579437
0.31	5.079039	5.638064	6.289875	7.054007	7.954518	9.021190
0.32	4.869585	5.390575	5.995840	6.702967	7.533406	8.513612
0.33	4.675331	5.161809	5.725066	6.380829	7.148198	8.050965
0.34	4.494838	4.949990	5.475180	6.084525	6.795200	7.628322

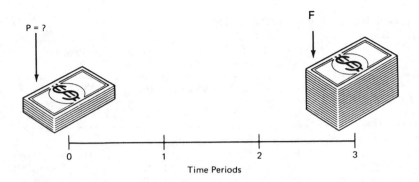

Fig. 4-2 Single payment present worth (P/F).

Uniform Series Compound Amount—F/A

The F/A factor is used to determine the amount F that an equal annual payment A will accumulate to in n years at i percent interest. If A (uniform annual payment) is known, and F (the future worth of these payments) is required, then Equation 4-6 is used.

$$F = A \times \frac{(1+i)^n - 1}{i} \tag{4-6}$$

$$F/A = \frac{(1+i)^n - 1}{i} \tag{4-7}$$

Uniform Series Present Worth—(P/A)

The P/A factor is used to determine the present amount P that can be paid by equal payments of A (uniform annual payment) at i percent interest, for n years. If A is known, and P is required, then Equation 4-8 is used.

$$P = A \times \frac{(1+i)^n - 1}{i(1+i)^n} \tag{4-8}$$

$$P/A = \frac{(1+i)^n - 1}{i(1+i)^n} \tag{4-9}$$

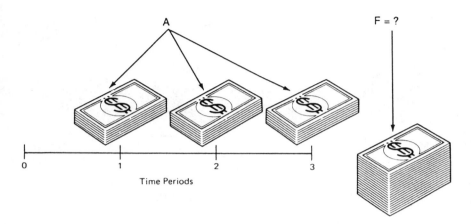

Fig. 4-3. Uniform series present worth (F/A).

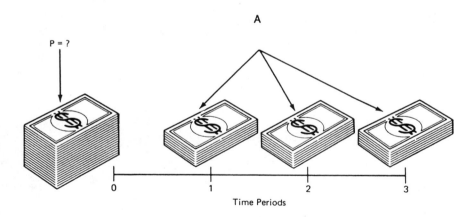

Fig. 4-4. Uniform series present worth (P/A).

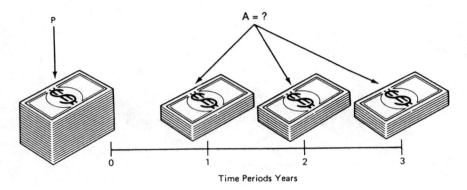

Fig. 4-5. Capital recovery (A/P).

Capital Recovery—A/P

The A/P factor is used to determine an annual payment A required to pay off a present amount P at i percent interest, for n years. If the present sum of money, P, spent today is known, and the uniform payment A needed to pay back P over a stated period of time is required, then Equation 4-10 is used

$$A = P \times \frac{i(1 + i)^n}{(1 + i)^n - 1}$$

(4-10)

$$A/P = \frac{i(1 + i)^n}{(1 + i)^n - 1}$$

(4-11)

Sinking Fund Payment—A/F

The A/F factor is used to determine the equal annual amount R that must be invested for n years at i percent interest in order to accumulate a specified future amount. If F (the future worth of a series of annual payments) is known, and A (value of those annual payments) is required, then Equation 4-12 is used.

$$A = F \times \frac{i}{(1+i)^n - 1} \qquad (4\text{-}12)$$

$$A/F = \frac{i}{(1+i)^n - 1} \qquad (4\text{-}13)$$

Gradient Present Worth—GPW

The GPW factor is used to determine the present amount P that can be paid by annual amounts A' which escalate at e percent, at i percent interest, for n years. If A' is known, and P is required, then Equation 4-14 is used. The GPW factor is a relatively new term which has gained in importance due to the impact of inflation.

$$P = A' \times (\text{GPW}) \, i_n \qquad (4\text{-}14)$$

$$P/A' = \text{GPW} = \frac{\dfrac{1+e}{1+i}\left[1 - \left(\dfrac{1+e}{1+i}\right)^n\right]}{1 - \dfrac{1+e}{1+i}} \qquad (4\text{-}15)$$

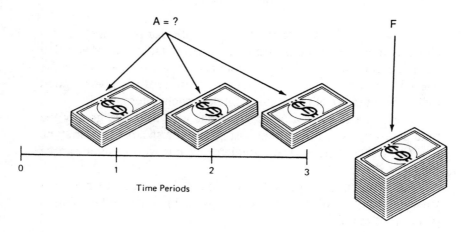

Fig. 4-6. Sinking fund payment (A/F).

The three most commonly used methods in life cycle costing are the annual cost, present worth and rate-of-return analysis.

In the present worth method a minimum rate of return (i) is stipulated. All future expenditures are converted to present values using the interest factors. The alternative with lowest effective first cost is the most desirable.

A similar procedure is implemented in the annual cost method. The difference is that the first cost is converted to an annual expenditure. The alternative with lowest effective annual cost is the most desirable.

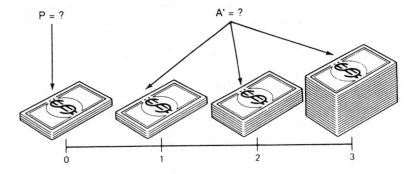

Fig. 4-7. Gradient present worth

In the rate-of-return method, a trial-and-error procedure is usually required. Interpolation from the interest tables can determine what rate of return (i) will give an interest factor which will make the overall cash flow balance. The rate-of-return analysis gives a good indication of the overall ranking of independent alternates.

The effect of escalation in fuel costs can influence greatly the final decision. When an annual cost grows at a steady rate it may be treated as a gradient and the gradient present worth factor can be used.

Special thanks are given to Rudolph R. Yanuck and Dr. Robert Brown for the use of their specially designed interest and escalation tables used in this text.

When life cycle costing is used to compare several alternatives the differences between costs are important. For example, if one alternate forces additional maintenance or an operating expense to occur, then these factors as well as energy costs need to be included. Remember, what was previously spent for the item to be replaced is irrelevant. The only factor to be considered is whether the new cost can be justified based on projected savings over its useful life.

THE JOB SIMULATION EXPERIENCE

Throughout the text you will experience job situations and problems. Each simulation experience is denoted by SIM. The answer will be given below the problem. Cover the answers, then you can "play the game."

SIM 4-1
An evaluation needs to be made to replace all 40-watt fluorescent lamps with a new lamp that saves 12 percent or 4.8 watts and gives the same output. The cost of each lamp is $2.80.

Assuming a rate of return before taxes of 25 percent is required, can the immediate replacement be justified? Hours of operation are 5800 and the lamp life is two years. Electricity costs 7.0¢/kWh.

ANSWER
$$A = 5800 \times 4.8 \times 0.070/1000 = \$1.94$$

$$A/P = 1.94/2.80 = .69$$

From Table 4-5 a rate of return of 25 percent is obtained. When analyzing energy conservation measures, never look at what was previously spent or the life remaining. Just determine if the new expenditure will pay for itself.

SIM 4-2
An electrical energy audit indicates electrical motor consumption is 4×10^6 kWh per year. By upgrading the motor spares with high efficiency motors a 10% savings can be realized. The additional cost for these motors is estimated at $80,000. Assuming an 8¢ per kWh energy charge and 20-

year life, is the expenditure justified based on a minimum rate of return of 20% before taxes? Solve the problem using the present worth, annual cost, and rate-of-return methods.

Analysis

Present Worth Method

	Alternate 1 Present Method	Alternate 2 Use High Efficiency Motor Spares
(1) First Cost (P)	—	$80,000
(2) Annual Cost (A)	$4 \times 10^6 \times .08$ = $320,000	$.9 \times \$320,000$ = $288,000
P/A (Table 4-4)	4.87	4.87
(2) A × 4.87 =	$1,558,400	$1,402,560
Present Worth	$1,558,400	$1,482,560
(1) + (3)		Choose Alternate with Lowest First Cost

Annual Cost Method

	Alternate 1	Alternate 2
(1) First Cost (P)	—	$80,000
(2) Annual Cost (A)	$320,000	$288,000
A/P (Table 4-4)	.2	.2
(3) P × .2	—	$16,000
Annual Cost	$320,000	$304,000
(2) + (3)		Choose Alternate with Lowest First Cost

Rate of Return Method

$$P = (\$320,000 - \$288,000)$$

$$P/A = \frac{80,000}{32,000} = 2.5$$

What value of i will make P/A = 2.5? $i = 40\%$ (Table 4-7).

SIM 4-3

Show the effect of 10 percent escalation on the rate of return analysis given the

Energy equipment investment   = $20,000
After tax savings   = $ 2,600
Equipment life (n)   = 15 years

ANSWER
Without escalation:

$$\frac{A}{P} = \frac{2,600}{20,000} = 0.13$$

From Table 4-1, the rate of return is 10 percent. With 10 percent escalation assumed:

$$\frac{P}{A} = \frac{20,000}{2,600} = 7.69$$

From Table 4-11, the rate of return is 21 percent.

Thus we see that taking into account a modest escalation rate can dramatically affect the justification of the project.

MAKING DECISIONS FOR ALTERNATE INVESTMENTS

There are several methods for determining which energy conservation alternative is the most economical. Probably the most familiar and trusted method is the annual cost method.

When evaluating replacement of processes or equipment *do not* consider what was previously spent. The decision will be based on whether the new process or equipment proves to save substantially enough in operating costs to justify the expenditure.

Equation 4-16 is used to convert the lump sum investment P into the annual cost. In the case where the asset has a value after the end of its useful life, the annual cost becomes:

$$AC = (P - L) * A/P + iL \qquad (4\text{-}16)$$

where
AC is the annual cost
L is the net sum of money that can be realized for a piece of equipment, over and above its removal cost, when it is returned at the end of the service life. *L is* referred to as the salvage value.

As a practical point, the salvage value is usually small and can be neglected, considering the accuracy of future costs. The annual cost technique can be implemented by using the following format:

	Alternate 1	Alternate 2
1. First cost (P)		
2. Estimated life (n)		
3. Estimated salvage value at end of life (L)		
4. Annual disbursements, including energy costs & maintenance (E)		
5. Minimum acceptable return *before* taxes (i)		
6. A/P n, i		
7. $(P - L) * $ A/P		
8. Li		
9. $AC = (P - L) * $ A/P $ + Li + E$		

Choose alternate with lowest AC

The alternative with the lowest annual cost is the desired choice.

SIM 4-4
A new water line must be constructed from an existing pumping station to a reservoir. Estimates of construction and pumping costs for each pipe size have been made.

Pipe Size	Estimated Construction Costs	Cost/Hour for Pumping
8"	$80,000	$4.00
10"	$100,000	$3.00
12"	$160,000	$1.50

The annual cost is based on a 16-year life and a desired return on investment, before taxes of 10 percent. Which is the most economical pipe size for pumping 4000 hours/year?

ANSWER

	8" Pipe	10" Pipe	12" Pipe
P	$80,000	$100,000	$160,000
n	16	16	16
E	16,000	12,000	6,000
i	10%	10%	10%
A/P = 0.127	—	—	—
(P − L) A/P	10,160	12,700	20,320
Li	—	—	—
AC	$26,160	$24,700 *(Choice)*	$26,320

DEPRECIATION, TAXES, AND THE TAX CREDIT

Depreciation

Depreciation affects the "accounting procedure" for determining profits and losses and the income tax of a company. In other words, for tax purposes the expenditure for an asset such as a pump or motor cannot be fully expensed in its first year. The original investment must be charged off for tax purposes over the useful life of the asset. A company usually wishes to expense an item as quickly as possible.

The Internal Revenue Service allows several methods for determining the annual depreciation rate.

Straight-Line Depreciation. The simplest method is referred to as a straight-line depreciation and is defined as:

$$D = \frac{P - L}{n}$$

(4-17)

where
D   is the annual depreciation rate
L   is the value of equipment at the end of its useful life, commonly referred to as salvage value
n   is the life of the equipment, which is determined by Internal Revenue Service guidelines
P   is the initial expenditure.

Sum-of-Years Digits. Another method is referred to as the sum-of-years digits. In this method the depreciation rate is determined by finding the sum of digits using the following formula,

$$N = n \frac{(n+1)}{2}$$ (4-18)

where *n is* the life of equipment.

Each year's depreciation rate is determined as follows.

First year $$D = \frac{n}{N} (P - L)$$ (4-19)

Second year $$D = \frac{n-1}{N} (P - L)$$ (4-20)

n year $$D = \frac{1}{N} (P - L)$$ (4-21)

Declining-Balance Depreciation. The declining-balance method allows for larger depreciation charges in the early years which is sometimes referred to as fast write-off.

The rate is calculated by taking a constant percentage of the declining undepreciated balance. The most common method used to calculate the declining balance is to predetermine the depreciation rate. Under certain circumstances a rate equal to 200 percent of the straight-line depreciation rate may be used. Under other circumstances the rate is limited to 1-1/2 or 1/4 times as great as straight-line depreciation. In this method the salvage value or undepreciated book value is established once the depreciation rate is pre-established.

To calculate the undepreciated book value, Equation 4-22 used.

$$D = 1 - \left(\frac{L}{P}\right)^{1/N}$$ (4-22)

where

D is the annual depreciation rate
L is the salvage value
P is the first cost.

The Tax Reform Act of 1986 (hereafter referred to as the "Act") represented true tax reform, as it made sweeping changes in many basic federal tax code provisions for both individuals and corporations. The Act has had significant impact on financing for cogeneration, alternative energy and energy efficiency transactions, due to substantial modifications in provisions concerning depreciation, investment and energy tax credits, tax-exempt financing, tax rates, the corporate minimum tax and tax shelters generally.

The Act lengthened the recovery periods for most depreciable assets. The Act also repealed the 10 percent investment tax credit ("ITC") for property placed in service on or after January 1, 1986, subject to the transition rules.

Tax Considerations

Tax-deductible expenses such as maintenance, energy, operating costs, insurance, and property taxes reduce the income subject to taxes.

For the after-tax life cycle analysis and payback analysis the actual incurred and annual savings is given as follows.

$$AS = (1 - I)E + ID \qquad (4\text{-}23)$$

where

AS *is* the yearly annual after-tax savings (excluding effect of tax credit)
E *is* the yearly annual energy savings (difference between original expenses and expenses after modification)
D *is* the annual depreciation rate is the income tax bracket.
I is the income tax bracket

Equation 4-23 takes into account that the yearly annual energy savings is partially offset by additional taxes which must be paid due to reduced operating expenses. On the other hand, the depreciation allowance reduces taxes directly.

After-Tax Analysis

To compute a rate of return which accounts for taxes, depreciation, escalation, and tax credits, a cash-flow analysis is usually required. This method analyzes all transactions including first and operating costs. To determine the after-tax rate of return a trial and error or computer analysis is required.

All money is converted to the present assuming an interest rate. The summation of all present dollars should equal zero when the correct interest rate is selected, as illustrated in Fig. 4-8.

This analysis can be made assuming a fuel escalation rate by using the gradient present worth interest of the present worth factor.

| | 1 | 2 | 3 | 4 | |
| | | | After Tax Savings (AS) | Single Payment Present Worth Factor | $(2+3) \times 4$ Present Worth |
Year	Investment	Tax Credit			
0	$-P$				$-P$
1		$+TC$	AS	P/F_1	$+P_1$
2			AS	P/F_2	P_2
3			AS	P/F_3	P_3
4			AS	P/F_4	P_4
Total					Σp

$$AS = (1 - I)\,E + ID$$
Trial and Error Solution:
Correct i when $\Sigma P = 0$

Fig. 4-8. Cash flow rate of return analysis.

SIM 4-5

Develop a set of curves that indicate the capital that can be invested to give a rate of return of 15 percent after taxes for each $1000 saved for the following conditions.

1.   The effect of escalation is not considered.
2.   A 5 percent fuel escalation is considered.
3.   A 10 percent fuel escalation is considered.
4.   A 14 percent fuel escalation is considered.
5.   A 20 percent fuel escalation is considered.

   Calculate for 5-, 10-, 15-, 20-year life.
   Assume straight-line depreciation over useful life, 34 percent income tax bracket, and no tax credit.

ANSWER $AS = (1 - I) E + ID$

$$I = 0.34, \qquad E = \$1000$$

$$AS = 660 + \frac{0.34P}{N}$$

   Thus, the after-tax savings (AS) are comprised of two components. The first component is a uniform series of $660 escalating at e percent/year. The second component is a uniform series of 0.34 P/N.
   Each component is treated individually and converted to present day values using the GPW factor and the P/A factor, respectively. The sum of these two present worth factors must equal P. In the case of no escalation, the formula is:

$$P = 660 * P/A + \frac{0.34P}{N} * P/A$$

In the case of escalation:

$$P = 660 \, GPW + \frac{0.34P}{N} * P/A$$

   Since there is only one unknown, the formulas can be readily solved. The results are indicated below.

	$N = 5$ $P	$N = 10$ $P	$N = 15$ $P	$N = 20$ $P
$e = 0$	2869	4000	4459	4648
$e = 10\%$	3753	6292	8165	9618
$e = 14\%$	4170	7598	10,676	13,567
$e = 20\%$	4871	10,146	16,353	23,918

Figure 4-9 illustrates the effects of escalation. This figure can be used as a quick way to determine after-tax economics of energy utilization expenditures.

SIM 4-6
It is desired to have an after-tax savings of 15 percent. Calculate the investment that can be justified if it is assumed that the fuel rate escalation should not be considered and the annual energy savings is $2000 with an equipment economic life of 15 years.
    Calculate the investment that can be justified in the above example, assuming a fuel rate escalation of 14%.

ANSWER
From Fig. 4-9, for each $1000 energy savings, an investment of $4400 is justified or $8800 for a $2000 savings when no fuel increase is accounted for.
    With a 14 percent fuel escalation rate an investment of $10,600 is justified for each $1000 energy savings, thus $21,200 can be justified for $2000 savings. Thus, a much higher expenditure is economically justifiable and will yield the same after-tax rate of return of 15 percent when a fuel escalation of 14 percent is considered.

IMPACT OF FUEL INFLATION
ON LIFE CYCLE COSTING

As illustrated by problem 4-5 a modest estimate of fuel inflation as a major impact on improving the rate of return on investment of the project. The problem facing the energy engineer is how to fore-

Figure 4-9. Effects of Escalation On Investment Requirements

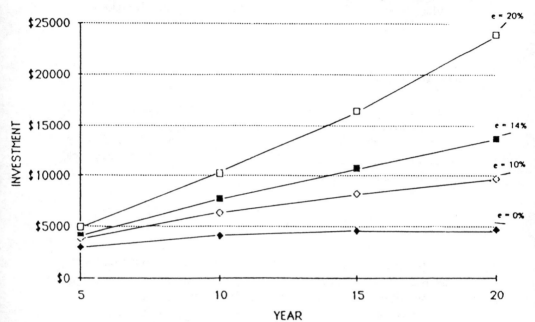

Note: Maximum investment in order to attain a 15% after-tax rate of return on investment for annual savings of $1000.

cast what the future of energy costs will be. All too often no fuel inflation is considered because of the difficulty of projecting the future. In making projections the following guidelines may be helpful:

• Is there a rate increase that can be forecast based on new nuclear generating capacity?

• What has been the historical rate increase for the facility? Even with fluctuations there are likely to be trends to follow.

• What events on a national or international level would impact on your costs? New state taxes, new production quotas by OPEC and other factors affecting your fuel prices.

- What do the experts say? Energy economists, forecasting services, and your local utility projections all should be taken into account.

SUMMARY OF LIFE-CYCLE COSTING

Always draw a cash flow diagram on a time basis scale. Show cash flow ins as positive and cash flow outs as negative.

In determining which interest formula to use, the following procedure may be helpful. First, put the symbols in two rows, one above the other as below:

$$\frac{PAF\ (unknown)}{PAF\ (known)}$$

The top represents the unknown values, and the bottom line represents the known. From information you have and desire, simply circle one of each line, and you have the correct factor.

For example, if you want to determine the annual saving "A" required when the cost of the energy device "P" is known, circle P on the bottom and A on the top. The factor A/P or capital recovery is required for this example. Table 4-13 summarizes the cash analysis for interest formulas.

GIVEN	FIND	USE
P	F	F/P
F	P	P/F
A	F	F/A
F	A	A/F
P	A	A/P
A	P	P/A

Table 4-13. Cash Analysis for Interest Formulas

EQUIPMENT LIFE

To estimate equipment lifefor life cycle cost analysis, Table 4-14 can be used.

COMPUTER ANALYSIS

The Alliance to Save Energy, 1925 K Street, NW, Suite 206, Washington, DC 20006, has introduced an investment analysis software package, ENVEST, which costs only $75 for 5-1/2-inch disks and $85 for 3-1/2-inch hard drive disks, and includes a 170-page user manual and 30 days of telephone support. The program can be run on an IBM PC, PCXT, with 256K ram.

Table 4-14. Equipment Service Life Statistics

				Percentiles		
Equipment Item	Mean	Median	Model(s)	25%	75%	N
UNITARY EQUIPMENT						
Room Air Conditioners						
(window or through-the-wall)	10	10	10	5	10	38
Unitary Air Conditioners						
1. Air-cooled—residential	14	15	15	8	20	29
(single package or split system						
2. Air cooled—commercial/industrial	15	15	15	10	20	40
(single package—through-the-wall						
or split system)						
3. Water cooled—electric	16	15	15-20	10	20	17
Unitary Heat Pumps						
1. Air source—residential	11	10	10	10	12.5	12

(Continued)

Equipment Item	Mean	Median	Model(s)	Percentiles 25%	75%	N
2. Air source—commercial/industrial (single package or split system)	15	15	15	11	15	13
3. Water source—comm./industrial	13	13	10	10	20	8
Computer Room Conditioners	18	15	15	15	20	23
ROOF TOP HVAC SYSTEMS Single Zone Heating, ventilating and cooling or cooling only	15	15	15	10	20	30
Multizone Heating, ventilating and cooling or cooling only	16	15	15	10	20	25
HEATING EQUIPMENT Boilers 1. Steam —steel watertube	30	26	40	20	40	30
—steel firetube	24	25	25	20	30	14
—cast iron	30	30	30	20	35	12
2. Hot water —steel watertube	24	23	20	20	27	12
—steel firetube	23	24	30	17	30	16
—cast iron	30	30	30	20	40	13
3. Electric	14	15	15	7	17	9
Burners Gas—forced and natural and oil-forced	22	20	20	17	27	58
Furnaces Gas or oil	18	20	20	12	20	35
Unit Heaters Gas or electric	14	13	10	10	20	28
Hotwater or steam	23	20	20	20	30	30

(Continued)

Equipment Item	Mean	Median	Model(s)	Percentiles 25%	75%	N
Radiant Heaters and Panels						
Electric heaters	11	10	10	5	25	6
Hot water or steam panels	26	25	20-25	20	30	7
AIR HANDLING AND TREATING EQUIPMENT						
Terminal Units						
1. Induction units	26	20	20	20	30	16
2. Fan coil	21	20	20	16	22	28
3. Diffusors, grilles and registers	35	27	20	20	50	26
4. Double duct mixing boxes— constant or variable air volume	21	20	20	15	30	20
5. Variable air volume (VAV) boxes single duct	24	20	20	20	30	7
Air Washers	20	17	30	10	30	6
Humidifiers	18	15	10	10	20	23
Ductwork	35	30	50	24	50	31
Dampers including actuators	15	20	20	15	30	20
Fans (supply or exhaust)						
1. Centrifugal—forward curve or backward inclined	27	25	20	20	40	43
2. Axial flow	23	20	20	10	30	16
3. Wall-mounted—propeller type	17	15	20	10	20	15
4. Ventilating—roof mounted	17	20	20	10	20	22
HEAT EXCHANGERS						
Coils						
1. DX	22	20	20	15	27	21
2. Water or steam	24	20	20	20	30	49
3. Electric	15	15	10-15-20	10	20	9
Shell and Tube	25	24	20	20	30	20
COOLING EQUIPMENT						
Reciprocating Compressors	18	20	20	12	20	7

(Continued)

Equipment Item	Mean	Median	Model(s)	Percentiles		N
				25%	75%	
Chillers —packaged—reciprocating	19	20	20	15	20	34
—centrifugal	25	23	20	20	30	28
—absorption	24	23	20	20	30	16
HEAT REJECTION EQUIPMENT						
Cooling Tower —metal—galvanized	18	20	20	10	20	33
—wood	22	20	20	15	27	25
—ceramic	33	34	20	20	5	6
Air-cooled Condenser	20	20	20	15	25	27
Evaporative Condenser	18	20	20	15	20	13
GENERAL COMPONENTS						
Insulation						
1. Preformed—block, molded, etc.	27	20	20	20	30	43
2. Blankets, batts	29	24	20	20	40	23
Pumps						
1. Circulating, base-mounted	19	20	20	13	24	37
2. Circulating, pipe-mounted	12	10	10-15	6	15	28
3. Sump and well	15	10	30	6	30	25
4. Condensate and receiver	18	15	15	10	25	25
Engines, Turbines, Motors						
1. Reciprocating engine	19	20	20	20	20	12
2. Turbines—steam	30	30	40	24	30	13
3. Electric motors	18	18	20	13	20	24
Motor Starters—across line or magnetic	19	17	20	10	30	34
Transformers						
Dry type or oil-filled	31	30	30	20	40	49
Controls and instrumentation						
1. Pneumatic	21	20	20	15	24	34
2. Electrical	17	16	20	10	20	24

(Continued)

Equipment Item	Mean	Median	Model(s)	Percentiles		N
				25%	75%	
3. Electronic	15	15	10-15	10	20	16
4. Automated (computer) building control systems	22	20	20-25	10	25	8
Valve Actuators						
1. Electric	16	14	10-20-30	5	25	18
2. Hydraulic	15	15	20	5	24	8
3. Pneumatic	18	20	20	10	25	26
4. Self-contained	14	10	5-20	5	24	9

5

The Heating, Ventilation And Air-Conditioning Audit

Energy audits of Heating, Ventilation and Air-Conditioning (HVAC) Systems is a very important portion of the overall program. HVAC standards such as ASHRAE 90.1 exist for defining energy-efficient systems in new construction. On the other hand, as of this writing no standards exist to define HVAC-efficient systems for existing buildings.

The purpose of this chapter is to highlight "low cost-no cost" areas that should be investigated in the HVAC Energy Audit. Portions of material used in this section and audit forms appearing in Chapter 15 are based upon two publications: "Guidelines for Saving Energy in Existing Buildings—Building Owners and Operators Manual," ECM-1; and "Engineers, Architects and Operators Manual," ECM-2. Both manuals were prepared for the government by Fred S. Dubin, Harold L. Mindell and Selwyn Bloome. The volumes were originally published by the U.S. Department of Commerce National Technical Information Service PB-249928 and PB-249929 and are available from Superintendent of Documents, U.S. Government Printing Office, Washington, D.C. 20402. Reference to ECM-1 and ECM-2 in the text refer to the original publication. The original document published is one of the most extensive works on energy conservation in existing buildings. The author expresses appreciation and credit to this work as one of the outstanding contributions in the energy audit field.

To use the short-cut methods described in this chapter and the chapter on building energy audits (Chapter 6), knowledge of local weather data is required. Chapter 15, Table 15-1 and Figures 15-1 through 15-5 should prove helpful.

In addition Chapter 15 contains various audit forms which can be modified to fit particular needs, such as those shown in Figures 15-6 through 15-24.

A more detailed engineering approach is sometimes required utilizing computer programs, discussed at the end of this chapter, or detailed manual calculations. For manual engineering calculations reference is made to "Cooling and Heating Load Calculations," available from: American Society of Heating, Refrigeration and Air Conditioning Engineers.

Complete engineering weather data can be found in Air Force Manual, "Facility Design and Planning in Engineering Weather Data," available from the Superintendent of Documents, Washington, DC 20402.

INDOOR AIR QUALITY (IAQ) STANDARD[1]

The most effective means to deal with an IAQ problem is to remove or minimize the pollutant source, when feasible. If not, dilution and filtration may be effective.

Dilution (increased ventilation) is to admit more outside air to the building. ASHRAE's former 1981 standard recommended 5 CFM/person outside air in an office environment. The new ASHRAE ventilation standard, 62-1989, now requires 20 CFM/person, if the prescriptive approach is used.

Increased ventilation will have an impact on building energy consumption. However, this cost need not be severe. If an airside economizer cycle is employed and the HVAC system is controlled to respond to IAQ loads as well as thermal loads,

---

[1] Indoor Air Quality: Problems and Cures. M. Black and W. Robertson. Presented at 13th World Energy Engineering Congress.

20 CFM/person need not be adhered to and the economizer hours will help attain air quality goals with energy savings at the same time. Incidentally, it was the energy cost of treating outside air that led to the 1981 standard. The superseded 1973 standard recommended 15-25 CFM/person.

Energy savings can also be realized by the use of improved filtration in lieu of the prescriptive 20 CFM/person approach. Improved filtration can occur at the air handler, in the supply and return ductwork, or in the spaces via self-contained units. Improved filtration can include enhancements such as ionization devices to neutralize airborne biological matter and to electrically charge fine particles, causing them to agglomerate and be more easily filtered.

Guidelines for IAQ pollutants are illustrated in Figure 5-1.

Specific methods are available for preventing or reducing IAQ concerns. These include:

1. Providing adequate and effective ventilation. This includes complying with the ASHRAE Standard 62-1989. The proper amount of outside air must be brought into the building and the air must be effectively distributed to the breathing level zone of the occupants. In addition, air intakes and exhaust systems must be designed so that polluted air is not brought into the building. Those activities which generate high loads of pollutants such as tobacco smoking areas and printing/graphic areas should be exhausted directly to the outside.

2. Insuring that safe, low emitting materials are used in new construction and remodeling activities. These include construction materials and furnishings such as wallboard, floor coverings, wallcoverings, paints, adhesives, duct lining, ceiling tiles, furniture, etc. These materials should be pre-tested or certified to be low emitting.

3. Enforcing a well-documented and scheduled HVAC operational plan which includes changing of filters, cleaning of air handling rooms, cleaning of condensate

Pollutant	Concentration	Remarks
Asbestos	0.2 fibers/cm^3	OSHA standard set in July, 1986.
	0.3 fibers/cm^3	OSHA action level requiring monitoring programs; typical background levels in outdoor ambient air in urban areas are 0.00007 fibers/cm^3.
Carbon Dioxide	1000 ppm	Japanese standard for buildings with floor space exceeding 3000 m^2 and HVAC system.
Carbon Monoxide	9 ppm	National Ambient Air Quality standard average of 8 hours.
Formaldehyde	0.1 ppm	ASHRAE recommended limit based on comfort criteria which should protect all but hypersensitive individuals.
	0.4 ppm	HUD standard for pressed wood products used in mobile homes, to prevent formaldehyde in indoor air from exceeding 0.4 ppm.
Nitrogen Dioxide	0.05 ppm	Annual National Ambient Air Quality standard.
Ozone	0.08 ppm	Level of concern in World Health Organization criteria documents.
	0.12 ppm	National Ambient Air Quality standard averaged over 1 hour.
Particulate	50 ug/m^3	National Ambient Air Quality standard annual geometric mean.
	150 ug/m^3	National Ambient Air Quality standard 24 hour average mean.
Radon	4 pCi/L	U.S. Environmental Protection Agency technologically achievable target level.
	8 pCi/L (0.04 WL)	Remedial action level recommended by the National Council on Radiation Protection and Measurements.
Termiticides	1.0 ug/m^3 (Aldrin) 5.0 ug/m^3 (Chlordane) 2.0 ug/m^3 () 10 ug/m^3 ()	Recommended by the National Academy of Sciences Committee on Toxicology.
Volatile Organic Compounds	1-5 mg/m^3	Lars Molhave study levels suspected of causing sick building syndrome symptoms in some individuals. U.S. EPA Guideline.

Figure 5-1. Guidelines for Some IAQ Pollutants

drip pans to discourage the growth of microbial debris, and assuring proper operational performance.

4. Removing or correcting existing sources of indoor pollutants. Special filtration, encapsulation and substitution are common techniques.

5. Educating the staff and building occupants concerning IAQ sources of pollutants and their effects, and control measures.

THE VENTILATION AUDIT

To accomplish an Energy Audit of the ventilation system the following steps can be followed:

1. Measure volume of air at the outdoor air intakes of the ventilation system. Record ventilation and fan motor nameplate data.

2. Determine local code requirements and compare against measurements.

3. Check if measured ventilation rates exceed code requirements.

To decrease CFM, the fan pulley can be changed. Two savings are derived from this change, namely:

• Brake horsepower of fan motor is reduced.
• Reduced heat loss during heating season.

To compute the savings Formulas 5-1 and 5-2 are used. Figure 5-2 can also be used to compute fan power savings as a result of air flow reduction.

$$\text{HP (Reduction)} = \text{HP} \times \left(\frac{\text{CFM (New)}}{\text{CFM (Old)}} \right)^3 \qquad \textit{Formula (5-1)}$$

$$Q \text{ (Saved)} = \frac{1.08 \text{ BTU}}{\text{HR}-\text{CFM}-\text{°F}} \times \text{CFM (Saved)} \times \Delta T \quad \textit{Formula (5-2)}$$

$$\text{KW} = \text{HP} \times .746/\eta \qquad \textit{Formula (5-3)}$$

Where
    HP = Motor Horsepower
  CFM = Cubic feet per minute
   ΔT = Average temperature gradient
   KW = Motor Kilowatts (K = 1000)
    η = Motor efficiency

In addition to reducing air flow during occupied periods, consideration should be given to shutting the system down during unoccupied hours.

If the space was cooled, additional savings will be achieved. The quantity of energy required to cool and dehumidify the ventilated air to indoor conditions is determined by the enthalpy difference between outdoor and indoor air. To compute the energy savings for the cooling season Figure 5-3 can be used.

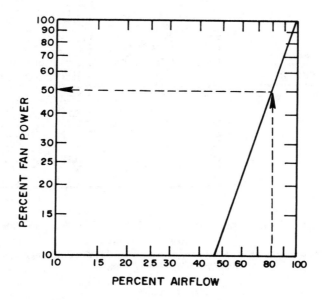

Figure 5-2. Decrease in Horsepower Accomplished By Reducing
Fan Speed (Based on Laws of Fan Performance)
(Source: NBS Handbook 115 Supplement 1)

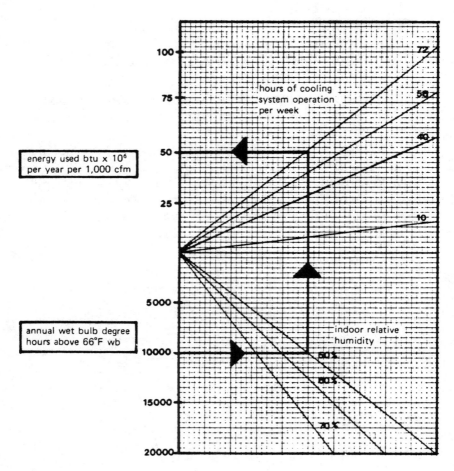

Figure 5-3. Yearly Energy Used Per 1,000 CFM to Maintain Various Humidity Conditions

(Source: Guidelines For Saving Energy in Existing Buildings—Building Owners and Operating Manual, ECM-1)

WE degree hours based on 12 Mos/Yr, 8 Hr/Day

Energy used is a function of the WE degree hours above the base of 66F, the RH maintained the No. of hours of controlled humidity. The base RH is 50% which is approximately 78F DB, 66 WB. The figure expresses the energy used per 1000 CFM of air conditioned or dehumidified.

SIM 5-1

An energy audit indicates ventilation in a storage area can be reduced from four to two changes per hour during the winter months– 240 days, 4200 degree-days.

Comment on the energy savings based on the following audit data:

- Building size: 20H X 150W X 100L
- Inside temperature: 70°F
- Motor Horsepower: 20 HP
- Nameplate Electrical Efficiency: .8
- Utility Costs: $4/10^6 BTU, 5¢ per KWH
- Hours of Operation 5760
- Boiler Efficiency = .65

ANALYSIS

Volume of Warehouse Area = 20 X 150 X 100 = 300,000 ft^3
Present Rate: 4 X 300,000 X 1/60 = 20,000 CFM
Reduced Rate = 2 X 300,000 X 1/60 = 10,000 CFM

Savings Due to Reduced Horsepower

Reduced Horsepower = 20 HP X $(2/4)^3$ = 2.5 HP
Savings Electricity = (20–2.5) X .746/.8 X 5¢ KW X 5760
    = $4,699.80

Savings Due to Reduced Heat Loss

Average ΔT = 4200 degree-days/240 days = 17.5°F
Average Outdoor = 65 – 17.5 = 47.5
Temperature during Heating Seasons

$$\text{Heat Removed} = \frac{1.08 \text{ BTU}}{\text{HR–CFM–°F}} \times \text{CFM } \Delta T =$$

1.08 (20,000 – 10,000) X 17.5 = 189,000 BTUH
Savings = 189,000 X $4/10^6 BTU/.65 X 5760 = $6698
Total Annual Savings = $4699 + 6698 = $11,397

SIM 5-2

For the building of SIM 5-1, compute the cooling savings resulting from reducing air changes per hour from 4 to 2.

Audit Data:
   Annual Wet Bulb Degree-Hours above 66°F = 8,000
   Relative Humidity = 50%
   Hours of Cooling System Operation per Week = 40
   Electricity Rate = 5¢ per KWH
   Refrigeration Consumption = .8 KW/Ton-Hr.

ANALYSIS

From Figure 5-3

Energy Used per year per 1000 CFM is 22.5×10^6 BTU

Energy Saved $= (20,000-10,000 \text{ CFM}) \times 22.5 \times 10^6$ BTU

$= 225 \times 10^6$ BTU/Yr.

$$\text{Savings} = \frac{225 \times 10^6 \text{ BTU/Yr}}{12,000 \text{ BTU/Ton-Hr}} \times .8 \text{ KW/Ton-Hr} \times 5¢/\text{KWH}$$

$= \$750/\text{Yr}$

THE TEMPERATURE AUDIT

The temperature audit should include the following:
- Determine indoor temperature settings for each space and season.
- Determine spaces which are unoccupied.
- Check if temperatures exceed "Recommended Temperature Standards," Figures 5-4 and 5-5.
- Implement setbacks by resetting thermostats manually, installing clocks or adjusting controls.
- Turn off cooling systems operated in summer during unoccupied hours.
- Experiment to determine optimum setback temperature.
- Lower temperature settings of occupied spaces based on "Recommended Temperature Standards," Figures 5-4 and 5-5.

	A Dry Bulb °F occupied hours maximum	B Dry Bulb °F unoccupied hours (set-back)	
1. OFFICE BUILDINGS, **RESIDENCIES, SCHOOLS**			
Offices, school rooms, residential spaces	68°	55°	
Corridors	62°	52°	
Dead Storage Closets	50°	50°	
Cafeterias	68°	50°	
Mechanical Equipment Rooms	55°	50°	
Occupied Storage Areas, Gymnasiums	55°	50°	
Auditoriums	68°	50°	
Computer Rooms	65°	As required	
Lobbies	65°	50°	
Doctor Offices	68°	58°	
Toilet Rooms	65°	55°	
Garages	Do not heat	Do not heat	
2. RETAIL STORES			
Department Stores	65°	55°	
Supermarkets	60°	50°	
Drug Stores	65°	55°	
Meat Markets	60°	50°	
Apparel (except dressing rms)	65°	55°	
Jewelry, Hardware, etc.	65°	55°	
Warehouses	55°	50°	
Docks and platforms	Do not heat	Do not heat	
3. RELIGIOUS BUILDINGS			
		24 Hrs or less	Greater than 24 Hrs
Meeting Rooms	68°	55°	50°
Halls of Worship	65°	55°	50°
All other spaces	As noted for office buildings	50°	40°

Source: Guidelines For Saving Energy in Existing Buildings—Building Owners
   and Operators Manual, ECM-1

Figure 5-4. Suggested Heating Season Indoor Temperatures

I. COMMERCIAL BUILDINGS	*Occupied Periods*	
	*Dry Bulb Temperature**	*Minimum Relative Humidity*
Offices	78°	55%
Corridors	Uncontrolled	Uncontrolled
Cafeterias	75°	55%
Auditoriums	78°	50%
Computer Rooms	75°	As needed
Lobbies	82°	60%
Doctor Offices	78°	55%
Toilet Rooms	80°	
Storage, Equipment Rooms	Uncontrolled	
Garages	Do Not Cool or Dehumidify.	

II. RETAIL STORES	*Occupied Periods*	
	Dry Bulb Temperature	*Relative Humidity*
Department Stores	80°	55%
Supermarkets	78°	55%
Drug Stores	80°	55%
Meat Markets	78°	55%
Apparel	80°	55%
Jewelry	80°	55%
Garages	Do Not Cool.	

* Except where terminal reheat systems are used. With terminal reheat systems the indoor space conditions should be maintained at lower levels to reduce the amount of reheat. If cooling energy is not required to maintain temperatures, 74°F would be recommended instead of 78°F.

Source: Guidelines For Saving Energy In Existing Buildings—Building Owners and Operators Manual, ECM-1

Figure 5-5. Suggested Indoor Temperature and Humidity Levels in the Cooling Season

Mandatory state and federal standards should be followed. Check that temperature requirements specified in OSHA are not violated. Changing temperatures in occupied periods could cause labor relations problems.

Other considerations in setting back temperatures during occupied hours include:

1. In spaces used for storage and which are mostly unoccupied, equipment and piping freeze protection is the main consideration.

2. Consider maintaining stairwell temperatures around 55°F in winter.

3. For areas where individuals commonly wear outdoor clothing such as stores, lower temperatures in winter.

SAVINGS AS A RESULT OF SETBACK

Figures 5-6 and 5-7 can be used to estimate savings as a result of setbacks for winter and summer respectively.

To use Figure 5-6:

1. Determine degree-days for location.

2. Calculate BTU/square foot/year used for heating.

3. Draw a line horizontally from specified degree-days to intersection of setback temperature. Extend line vertically and proceed along sloped lines as illustrated in the figure.

4. Draw a line horizontally from BTU/square foot/year until it intersects sloped line. Proceed vertically and read BTU/square foot/year savings on upper horizontal axis.

To use Figure 5-7:

1. Add the BTU/hour/1000 CFM for each temperature starting.

2. Start with one temperature above original set point.

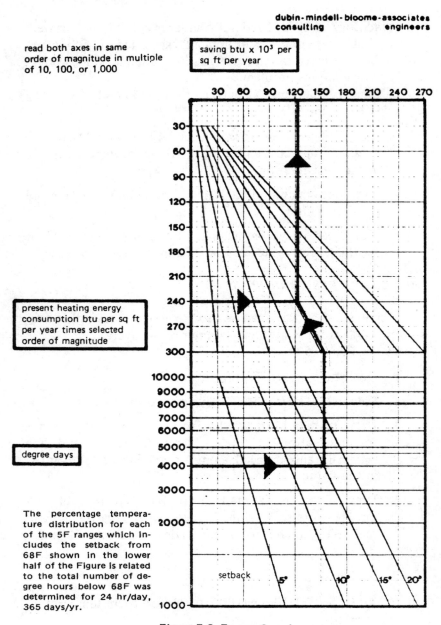

read both axes in same
order of magnitude in multiple
of 10, 100, or 1,000

saving btu x 10³ per
sq ft per year

dubin - mindell - bloome - associates
consulting   engineers

present heating energy
consumption btu per sq ft
per year times selected
order of magnitude

degree days

The percentage tempera-
ture distribution for each
of the 5F ranges which includes the setback from
68F shown in the lower
half of the Figure is related
to the total number of degree hours below 68F was
determined for 24 hr/day,
365 days/yr.

Figure 5-6. Energy Saved
(Source: Guidelines For Saving Energy In Existing Buildings—
Building Owners and Operators Manual, ECM-1)

3. Add contributions for each set point temperature until new setting is reached.

Relative Humidity	50%	60%	70%
Dry Bulb Temperature	BTU/Hour/1000 CFM		
72°F	0	0	0
73°F	2,700	2,433	3,000
74°F	2,657	2,400	3,257
75°F	3,000	2,572	3,000
76°F	3,000	2,572	3,000
77°F	3,000	2,572	3,429
78°F	3,000	2,572	3,429

Figure 5-7. Effect of Raising Dry Bulb Temperature
(Source: Guidelines For Saving Energy In Existing Buildings--
Building Owners and Operators Manual, ECM-1)

SIM 5-3

An energy audit indicates that the temperature of the building can be set back 20°F during unoccupied hours.
Comment on the energy savings based on the following audit data:
Heating Degree-Days = 6,000
Present Heating Consumption =
    60,000 BTU/square foot/year
Floor Area = 100,000 square feet
Utility Cost = $4/$10^6$ BTU
Boiler Efficiency = .65

ANALYSIS

From Figure 5-6:
Energy Savings is 36,000 BTU/square foot/year
Savings = 36,000 X $4/$10^6$ X 100,000/.65 = $22,153

SIM 5-4

An energy audit indicates an indoor dry bulb temperature in summer of 73°F. It is determined to raise the set point to 78°F. Comment on the energy savings based on the following audit data:

Total outdoor air	15,000 CFM
Relative Humidity	50%
Hours of Operation	40
Cooling Season	20 weeks/year
Annual W B degree-hours above 66°F, WB	6,000

ANALYSIS

From Figure 5-7, raising the temperature from 73°F to 78°F, a total savings of the following will occur:

74	2,656
75	3,000
76	3,000
77	3,000
78	3,000

Total.................14,656 BTU/Hour/1000 CFM
Savings: 14,656 BTU/Hour/1000 CFM X 15,000 CFM X $4/10^6 BTU/.65 X 40 Hours/Week X 20 Weeks/Year = $1,082 per year.

THE HUMIDITY AUDIT

Desired relative humidity requirements are achieved by vaporizing water into the dry ventilating air. Approximately 1000 BTUs are required to vaporize each pound of water. To save energy, humidification systems should not be used during unoccupied hours. Most humidification systems are used to maintain the comfort and health of occupants, to prevent cracking of wood, and to preserve materials. In lieu of specific standards it is suggested that 20% relative humidity be maintained in all spaces occupied more than four hours per day. If static shocks or complaints arise, increase the humidity levels in 5% increments until the appropriate level for each area is determined. Figure 5-8 can be used to estimate the savings in winter as a result of lowering the relative humidity requirements.

SIM 5-5

An energy audit of the humidification requirements indicated the following data:

Outdoor air rate plus infiltration	10,000 CFM
Annual Wet Bulb Degree-Hours	
Below 54°WB and 68°F	65,000
Cost of Fuel	$4 per million BTU
Boiler Efficiency	.65
Type	Department Store 112 hours occupancy/ week

Determine the savings as a result of lowering the relative humidity of the building from 50% to 30% during the heating season.

ANALYSIS

From Figure 5-8:

Energy used at 50 RH = 65 X 10^6 BTU/Yr per 1000 CFM
Energy used at 30 RH = 35 X 10^6 BTU/Yr per 1000 CFM
Energy saved = (65–35) X 10^6 X 10 = 300 X 10^6 BTU/Yr
Savings = 300 X 10^6 X $4/$10^6$ /.65 = $1,846
In the case of the cooling season check to determine if levels are consistent with Figure 5-5. Higher levels of humidification than required during the cooling season waste energy. Figure 5-9 should be used to estimate savings as a result of maintaining a higher RH level.

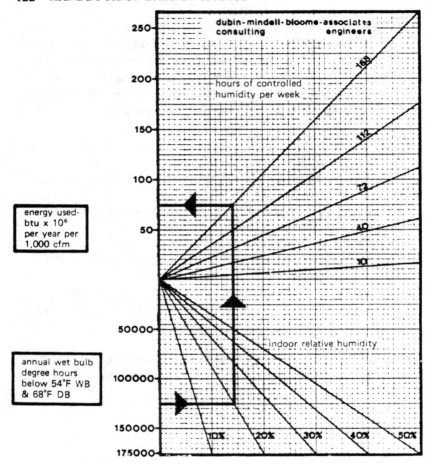

Figure 5-8. Yearly Energy Used Per 1,000 CFM to Maintain Various Humidity Conditions

(Source: Guidelines For Saving Energy In Existing Buildings—
Building Owners and Operators Manual, ECM-1)

WE degree hours based on 24 hours/day, October—April.
Base indoor condition for figure is DB=68F, WB=54F, RH=40%.
Energy used is a function of the WB degree hours below the base conditions, the RH maintained and the number of hours of controlled humidity. The figure expresses the energy used per 1000 cfm of air conditioned or humidified.

An analysis of the total heat content of air in the range under consideration indicates an average total heat variation of 0.522 Btu/lb for each degree WB change. Utilizing the specific heat of air, this can be further broken down to 0.24 Btu/lb sensible heat and 0.282 Btu/lb latent heat. 1000 cfm is equal to 4286 lb/hr and since we are concerned with latent heat only, each degree F WB hour is equal to 4286 × 0.282 or 1208 Btu. Further investigation of the relationship between WB temperature, DB temperature, and total heat shows that latent heat varies directly with RH at constant DB temperature. The lower section of the figure shows this proportional relationship around the base of 40% RH. The upper section proportions the hours of system operation with 168 hr/wk being 100%.

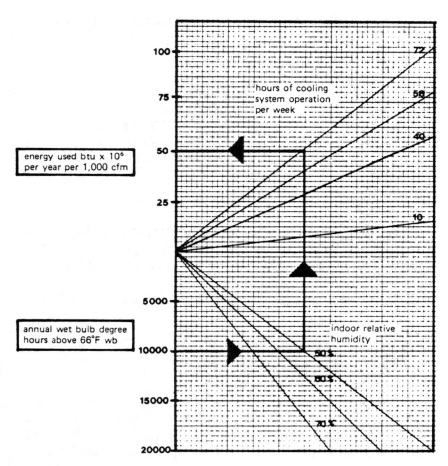

Figure 5-9. Yearly Energy Used Per 1,000 CFM to Maintain Various Humidity Conditions
(Source: Guidelines For Saving Energy In Existing Buildings—Building Owners and Operators Manual, ECM-1)

WE degree hours based on 12 Mos/Yr, 8 Hr/Day

Energy used is a function of the WB degree hours above the base of 66F, the RH maintained the No. of hours of controlled humidity. The base RH is 50% which is approximately 78F DB, 66F WB. The figure expresses the energy used per 1000 cfm of air conditioned or dehumidified.

SIM 5-6

Determine the savings based on increasing the relative humidity from 50% to 70% based on the audit data:

Annual Wet Bulb Degree-Hours above 66°F	8,000
Operation per week	40 hours
Outside CFM	20,000
Refrigeration consumptions	.8 KW/Ton Hour
Electric rate	5¢ per KWH

ANALYSIS

From Figure 5-9:

Energy used at 50% RH   22.5 X 10^6 BTU/Yr per 1000 CFM
Energy used at 70% RH   16 X 10^6 BTU/Yr per 1000 CFM
Energy saved = (22.5−16) X 10^6 X 20 = 130 X 10^6 BTU/Yr

$$\text{Savings} = \frac{130 \times 10^6}{12,000 \text{ BTU/Ton-Hr}} \times .8 \text{ KW/Ton-Hr} \times 5¢/\text{KWH}$$
$$= \$433/\text{Yr}$$

COMPUTER PROGRAM ANALYSIS

Computer program analysis is a very important design tool in the energy audit process. Manual load calculations are based on steady-state conditions. These calculations are usually based on maximum or minimum conditions and give reasonable indications of equipment size. They do not however indicate how the system will perform. Probably the greatest opportunity for savings exists under part-load conditions.

Computer programs simulate energy consumption based on stored weather data; this enables a comprehensive month-by-month energy report to determine the optimum system performance. The total system can be analyzed including lighting, HVAC, and building envelope. Thus alternatives may be investigated with all parameters considered.

ENERGY RECOVERY SYSTEMS

The HVAC Energy Audit should analyze opportunities for recovering energy. To recover heat from exhausts, several devices can be used including the heat wheel, air-to-air heat exchanger, heat pipe and coil run-around cycle. Examples of these systems and devices are illustrated in Figures 5-10 and 5-11.

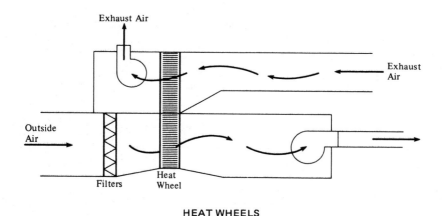

HEAT WHEELS

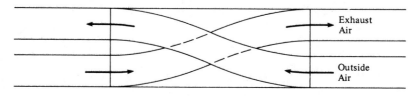

AIR-TO-AIR HEAT PIPES AND EXCHANGERS

Figure 5-10. HVAC Heat Recovery

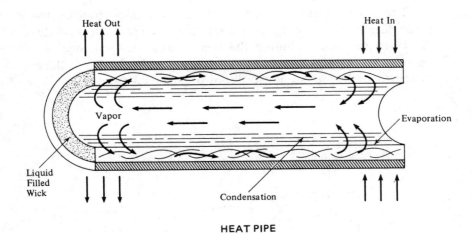

Heat Out

Heat In

Vapor

Evaporation

Liquid
Filled
Wick

Condensation

HEAT PIPE

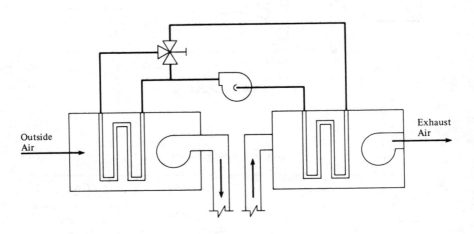

Outside
Air

Exhaust
Air

COIL-RUN-AROUND CYCLE

Figure 5-11. HVAC Heat Recovery

HEAT WHEELS

Heat wheels are motor-driven devices packed with heat-absorbing material such as a ceramic. As the device turns by means of a motor, heat is transferred from one duct to another.

AIR-TO-AIR HEAT EXCHANGER

The air-to-air heat exchanger consists of an open-ended steel box which is compartmentalized into multiple narrow channels. Each passage carries exhaust air alternating with make-up air. Energy is transmitted by means of conduction through the walls.

HEAT PIPES

A heat pipe is installed through adjacent walls of inlet and outlet ducts; it consists of a short length of copper tubing sealed at both ends. Inside is a porous cylindrical wick and a charge of refrigerant. Its operation is based on a temperature difference between the ends of the pipe, which causes the liquid in the wick to migrate to the warmer end to evaporate and absorb heat. When the refrigerant vapor returns through the hollow center of the wick to the cooler end, it gives up heat, condenses, and the cycle is repeated.

COIL RUN-AROUND CYCLE

The coil run-around cycle transfers energy from the exhaust stream to the make-up stream continuously circulating a heat transfer medium, such as ethylene glycol fluid, between the two coils in the ducts.

In winter, the warm exhaust air passes through the exhaust coils and transfers heat to the ethylene glycol fluid. The fluid is pumped to the make-up air coil where it preheats the incoming air. The system is most efficient in winter operation, but some recovery is possible during the summer.

HEAT FROM LIGHTING SYSTEMS

Heat dissipated by lighting fixtures which is recovered will reduce air-conditioning loads, will produce up to 13 percent more light output for the same energy input, and can be used as a source of hot air. Two typical recovery schemes are illustrated in Figure 5-12. In the total return system, all of the air is returned through the luminaires. In the bleed-off system, only a portion is drawn through the lighting fixtures. The system is usually used in applications requiring high ventilation rates.

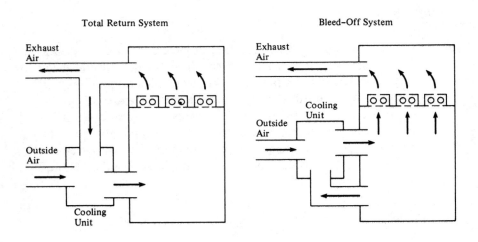

Figure 5-12. Recovery from Lighting Fixtures

SIM 5-7

Roof-mounted, air-cooled condensers are traditionally used to cool the gas from refrigeration equipment. Comment on how the system diagrammed below can be made more efficiently.

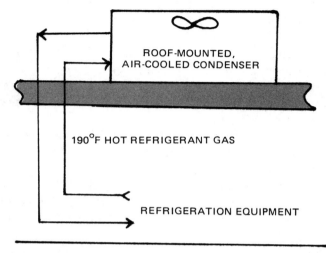

ROOF-MOUNTED,
AIR-COOLED CONDENSER

190°F HOT REFRIGERANT GAS

REFRIGERATION EQUIPMENT

ANALYSIS

This example illustrates a retrofit installation where heat is recovered by the addition of a heat exchanger to recapture the energy which was previously dissipated to the atmosphere. This energy can be used to preheat the domestic water supply for various processes.

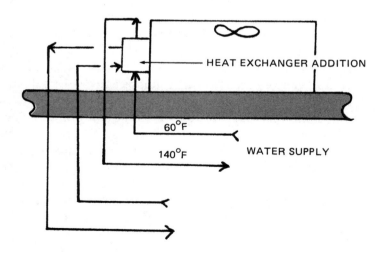

HEAT EXCHANGER ADDITION

60°F

140°F   WATER SUPPLY

ECONOMIZER CYCLE

In addition to heat recovery opportunities, the audit should uncover system modifications which will save energy such as the economizer cycle. The economizer cycle uses outside air as the cooling source when it is cold enough. There are two suitable economizer systems:

1. System monitors and responds to dry-bulb temperature only. It is suitable where wet-bulb degree-hours are less than 8000 per year.
2. System monitors and responds to the WB and DB temperatures (enthalpy), and is most effective and economic in locations which experience more than 8000 WB degree-hours.

SYSTEM 1 – ECONOMIZER CYCLE COOLING

Provide controls, dampers and interlocks to achieve the following control sequence:

a) When the outdoor air DB temperature is lower than the supply air DB temperature required to meet the cooling load, turn off the compressor and chilled water pumps, and position outdoor air–return air–exhaust air dampers to attain the required supply air temperature.

b) When the outdoor air DB temperature is higher than the supply air temperature required to meet the loads, but is lower than the return air temperature, energize the compressors and chilled water pumps and position dampers for 100% outdoor air.

c) Use minimum outdoor air whenever the outdoor dry-bulb temperature exceeds the return air DB temperature.

d) Whenever the relative humidity in the space drops below desired levels and more energy is consumed to raise the RH than is saved by the economizer system, consider using refrigeration in place of economizer cooling. This condition may exist in very cold climates and must be analyzed in detail.

SYSTEM 2 – ENTHALPY CYCLE COOLING

Provide the equipment, controls, dampers and interlocks to achieve the following control sequence:

The four conditions listed for system 1 above are similar for this system with the exception that enthalpy conditions are measured rather than dry-bulb conditions.

If changes to outside air intake are contemplated, take careful note of all codes bearing on ventilation requirements. Fire and safety codes must also be observed.

APPLICATIONS OF SYSTEMS 1 AND 2

- Single duct, constant volume systems
- Variable volume air systems
- Induction systems
- Terminal reheat systems, dual duct systems, and multi-zone systems.

Economizer and Enthalpy systems are less effective if used in conjunction with heat-recovery systems. Trade-offs should be analyzed.

TEST AND BALANCE CONSIDERATIONS

Probably the biggest overlooked low-cost energy audit requirement is a thorough Test, Balance and Adjust Program. In essence the audit should include the following steps:

1. *Test*–Quantitative determination of conditions within the system boundary, including flow rates, temperature and humidity measurements, pressures, etc.
2. *Balance*–Balance the system for required distribution of flows by manipulation of dampers and valves.
3. *Adjust*–Control instrument settings, regulating devices, control sequences should be adjusted for required flow patterns.

In essence the above program checks the designer's intent against actual performance and balances and adjusts the system for peak performance.

Several sources outlining Test and Balance Procedures are:

- Construction Specifications Institute (CSI), which offers a specification series that includes a guide specification entitled "Testing and Balancing of Environmental Systems."
- Associated Air Balance Council (AABC), the certifying body of independent agencies.
- National Environmental Balancing Bureau (NEBB), sponsored jointly by the Mechanical Contractors Association of America and the Sheet Metal and Air Conditioning Contractors National Association as the certifying body of the installing contractors' subsidiaries.

6

The Building System Energy Audit

The Building System Energy Audit (BSEA) requires gathering the following data:
1. Building characteristics and construction
   a. Window characteristics
   b. Openings and major leaks
2. Insulation status

BUILDING DYNAMICS

The building experiences heat gains and heat losses depending on whether the cooling or heating system is present, as illustrated in Figures 6-1 and 6-2. Only when the total season is considered in conjunction with lighting and heating, ventilation and air-conditioning (HVAC) can the energy choice be decided.

Many of the audits discussed in this chapter apply the principle of reducing the heat load or gain of the building. Thus the internal HVAC load would decrease. A caution should be made that without a detailed engineering analysis, a computer simulation, an oversimplification may lead to a wrong conclusion. The weather data for your area and the effect of the total system should not be overlooked.

In order to use the methods described in this chapter, weather data in Chapter 15, Table 15-1 and Figures 15-1 through 15-5 can be used. Figure 15-14 illustrates an energy audit form for a building that may be modified to suit your particular needs.

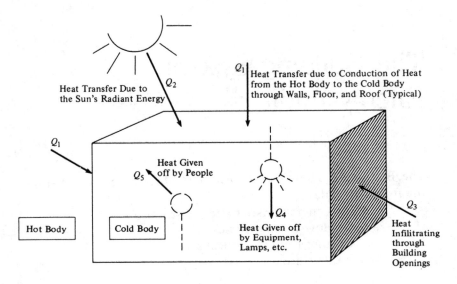

Heat Gain = $Q_1 + Q_2 + Q_3 + Q_4 + Q_5$

Figure 6-1. Heat Gain of a Building

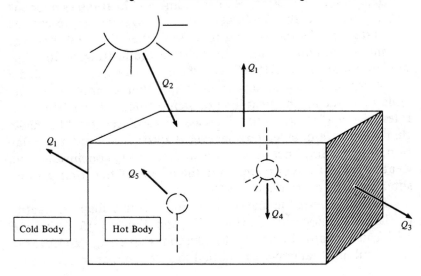

Heat Loss = $Q_1 + Q_3 - Q_2 - Q_4 - Q_5$

Figure 6-2. Heat Loss of a Building

SIM 6-1

Comment on the effect to the overall heat balance by adding skylights to the roof.

ANALYSIS

The effect of adding skylights will influence the overall energy balance in several ways.

1. The illumination from skylights will decrease the need for lighting systems. As an example a building with 6% coverage with skylights may receive ample illumination to turn off the lighting systems most daylight hours.
2. The solar heat gain factor is increased and if the building is air-conditioned more tons and more energy are required.
3. The excess solar heat gain during the winter months may decrease heating loads.

A detailed analysis is available from: Architectural Aluminum Manufacturers Association, 35 East Wacker Drive, Chicago, Illinois 60601 (312) 782-8256—"Voluntary Standard Procedure for Calculating Skylite Annual Energy Balance."

BUILDING CHARACTERISTICS AND CONSTRUCTION

The BSEA should record for each space the size, physical characteristics, hours of operation and function. The assorted materials of construction, windows, doors, holes, percentage glass, etc. should also be recorded.

INFILTRATION

Leakage or infiltration of air into a building is similar to the effect of additional ventilation. Unlike ventilation it cannot be controlled or turned off at night. It is the result of cracks, openings around windows and doors, and access openings. Infiltration is also induced into the building to replace exhaust air unless the

HVAC balances the exhaust. Wind velocity increases infiltration and stack effects are potential problems.

A handy formula which relates ventilation or infiltration rates to heat flow is Formula 6-1.

$$Q = 1.08 \times CFM \times \Delta T \qquad \textit{Formula (6-1)}$$

Where:

   Q is heat removal, BTU/Hr
   CFM is ventilation or infiltration rate, cubic feet per minute
   ΔT is the allowable heat rise.

Heat losses and gains from openings can significantly waste energy. All openings should be noted in the BSEA. Figure 6-3 illustrates the effect of the door size and time opened on the average annual heat loss. The graph is based upon a six-month heating season (mid-October to mid-April) and an average wind velocity of 4 mph. It is assumed that the heated building is maintained at 65°F. To adjust Figure 6-3 for different conditions use Formula 6-2.

$$Q = Q_1 \times \frac{d}{5} \times \frac{65 - T}{13} \qquad \textit{Formula (6-2)}$$

Where:

   Q is the adjusted heat loss, BTU/year
   Q is the heat loss from Figure 6-3
   d is the days of operation
   T is the average ambient temperature during the heating season, °F.

If the space was air-conditioned there would be an additional savings during the cooling season.

To reduce heat loss for operating doors, the installation of vinyl strips (see Figure 6-4) is sometimes used. This type of strip is approximately 90% efficient in reducing heat losses. The problem in using the strip is obtaining operator acceptance. Operators may feel these strips interfere with operations or cause a safety problem since vision through the access way is reduced.

An alternate method to reduce infiltration losses through access doors is to provide an air curtain.

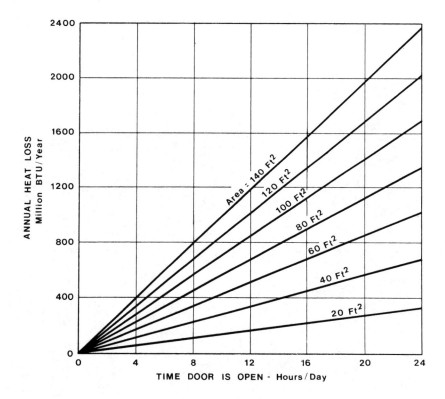

Figure 6-3. Annual Heat Loss from Doors
(Source: Georgia Tech Experiment Station)

SIM 6-2

An energy audit of a building indicated that the warehouse is maintained at 65°F during winter and has three 10 ft X 10 ft forklift doors. The warehouse is used 24 hours, 6 days per week, and the doors are open 8 hours per day. The average ambient temperature during the heating season is 48°F. Comment on adding vinyl strips (installed cost $2,000) which are 90% efficient, given the cost of heating fuel is $4/million BTU with a boiler efficiency of .65.

Figure 6-4. Installation of Vinyl Strips on Forklift Door

(Photographs courtesy of MetalGlas Products, Inc.)

ANALYSIS

From Figure 6-3, $Q = 600 \times 10^{12}$ BTU/Year

Therefore $Q = 600 \times 10^6 \times \frac{6}{5} \times \frac{65-48}{13} =$

941 million BTU/Yr per door

$Q = 3 \times .9 \times 941 \times \$4/\text{million BTU} + 65 = \$15.635$

Since the payback period before taxes is less than one year, the investment seems justified.

To estimate infiltration through windows Table 6-1 and Figure 6-5 may be used. This data also includes another estimating tool for determining infiltration through doors.

To compute the energy saved based on reducing the infiltration rates, Figure 6-6 and 6-7 are used for the heating and cooling seasons respectively.

SIM 6-3

An energy audit survey indicates 300 windows, poorly fitted wood sash, in a building which are not weatherstripped. Comment on the savings for weatherstripping given the following:

Data: Window size 54″ X 96″
        Degree-days = 8,000
        Cost of heating = $4/10^6$ BTU
        One-half the windows face the wind at any one time
        Hours of occupation = 5760
        Wet-Bulb Degree-hours = 2,000 greater than 66°F
        Wind velocity summer 10 MPH
        Refrigeration consumption = .8 KW/Ton-Hr
        Electric rate = 5¢ KWH
        Hours of operation = 72
        Indoor temperature winter 68°F
        RH summer 50%
        Boiler efficiency = .65

ANALYSIS

Area of windows $= \dfrac{54 \times 96}{144} = 36$ ft^2 per window

Coefficients from Table 6-1
With No Weatherstripping 1.52
With Weatherstripping   .47
Infiltration before = 36 X 300/2 X 1.52 = 8208 CFM
Infiltration after weatherstripping = 36 X 300/2 X .47 =
$$2538 \text{ CFM}$$
Savings with weatherstripping = 8208 - 2538 = 5670
From Figure 6-6 Q = 100 X 10^6 BTU/Year/1000 CFM
Savings during winter = 5.67 X 100 X 10^6 X $4/$10^6$/.65 =
$$\$3489$$
From Figure 6-7 Q = 10 X 10^6 BTU/Year/1000 CFM

Savings during summer (at 10 MPH wind velocity)
Savings summer = 5.67 X 10 X 10^6/12,000 X .8 X .05 X
$$10/15 = \$125.00$$

Total savings = $3614 per year

REDUCING INFILTRATION

In addition to weatherstripping, several key areas should not be overlooked in reducing infiltration losses.

Vertical shafts, such as stairwells, should be isolated as illustrated in Figure 6-8. Always check with fire codes before modifying building egress.

Poor quality outdoor air dampers are another source of excess infiltration. Dampers of this nature do not allow for accurate control and positive closure. Replacement with good quality opposed-blade dampers with seals at the blade edges and ends will reduce infiltration losses. (See Figure 6-9.)

The third area is to check exhaust hoods such as those used in kitchens and process equipment. Large open hoods are usually required to maintain a satisfactory capture velocity to remove fumes, smoke, etc. These hoods remove large volumes of air. The air is made up through the HVAC system which heats it up in winter and cools and dehumidifies it in summer. Several areas should be checked to reduce infiltration from hoods.

• Minimum capture velocity to remove contaminants.

Table 6-1. Infiltration Through Windows and Doors — Winter*

15 MPH Wind Velocity†

DOUBLE HUNG WINDOWS ON WINDWARD SIDE‡

| | CFM PER SQ FT AREA | | | | | |
| | Small — 30" X 72" | | | Large — 54" X 96" | | |
DESCRIPTION	No W-Strip	W-Strip	Storm Sash	No W-Strip	W-Strip	Storm Sash
Average Wood Sash	.85	.52	.42	.53	.33	.26
Poorly Fitted Wood Sash	2.4	.74	1.2	1.52	.47	.74
Metal Sash	1.60	.69	.80	1.01	.44	.50

NOTE: W-Strip denotes weatherstrip.

CASEMENT TYPE WINDOWS ON WINDWARD SIDE‡

| | CFM PER SQ FT AREA | | | | | | | | | |
| | Percent Ventilated Area | | | | | | | | | |
DESCRIPTION	0%	25%	33%	40%	45%	50%	60%	66%	75%	100%
Rolled Section—Steel Sash										
Industrial Pivoted	.65	1.44	—	1.98	—	1.1	—	2.9	—	5.2
Architectural Projected	—	.78	—	—	—	—	1.48	—	—	—
Residential	—	—	.56	—	—	.98	—	.63	—	1.26
Heavy Projected	—	—	—	—	.45	—	—	—	.78	—
Hollow Metal—Vertically Pivoted	.54	1.19	—	1.64	—	—	—	2.4	—	4.3

DOORS ON ONE OR ADJACENT WINDWARD SIDES‡

DESCRIPTION	Infrequent Use	CFM PER SQ FT AREA**			
			Average Use		
		1 & 2 Story Building	Tall Building (ft)		
			50	100	200
Revolving Door	1.6	10.5	12.6	14.2	17.3
Glass Door—(3/16" Crack)	9.0	30.0	36.0	40.5	49.5
Wood Door 3'7"	2.0	13.0	15.5	17.5	21.5
Small Factory Door	1.5	13.0			
Garage & Shipping Room Door	4.0	9.0			
Ramp Garage Door	4.0	13.5			

* All values are based on the wind blowing directly at the wind or door. When the prevailing wind direction is oblique to the window or doors, multiply the above values by 0.60 and use the total window and door area on the windward side(s).

† Based on a wind velocity at 15 mph. For design wind velocities different from the base, multiply the table values by the ratio of velocities.

‡ Stack effect in tall buildings may also cause infiltration on the leeward side. To evaluate this, determine the equivalent velocity (V_e) and subtract the design velocity (V). The equivalent velocity is:

$$V_e = \sqrt{V^2 - 1.75a} \quad \text{(upper section)}$$
$$V_e = \sqrt{V^2 - 1.75b} \quad \text{(lower section)}$$

Where a and b are the distances above and below the mid-height of the building, respectively, in ft.
Multiply the table values by the ratio $(V_e - V)/15$ for doors and one-half of the windows on the leeward side of the building. (Use values under "1 & 2 Story Building" for doors on leeward side of tall buildings.)

** Doors on opposite sides increase the above values 25%. Vestibules may decrease the infiltration as much as 30% when door usage is light. If door usage is heavy, the vestibule is of little value in reducing infiltration. Heat added to the vestibule will help maintain room temperature near the door.

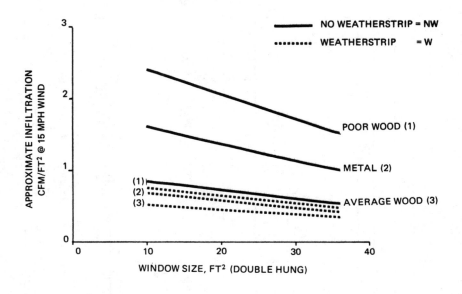

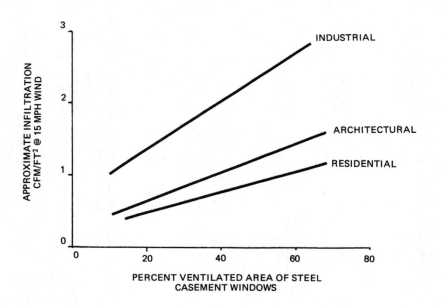

Figure 6-5. Infiltration Through Windows and Door — Winter
(Source: Instructions For Energy Auditors, Vol. II)

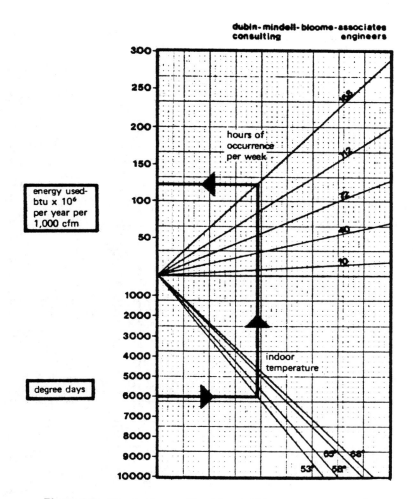

Figure 6-6. Yearly Energy Used Per 1,000 CFM Outdoor Air
(Source: Guidelines For Saving Energy In Existing Buildings—
Building Owners and Operators Manual, ECM-1)

Energy used is a function of the number of degree days, indoor temperature and the number of hours that temperature is maintained and is expressed as the energy used per 1000 cfm of air conditioned.

The energy used per year was determined as follows:

Btu/yr = (1000 cfm) (Degree Days/yr) (24 hr/day 1.08)*

Since degree days are base 65F, the other temperatures in the lower section of the figure are directly proportional to the 65F line. The upper section proportions the hours of system operation with 168 hr/wk being 100%.

*1.08 is a factor which incorporates specific heat, specific volume, and time.

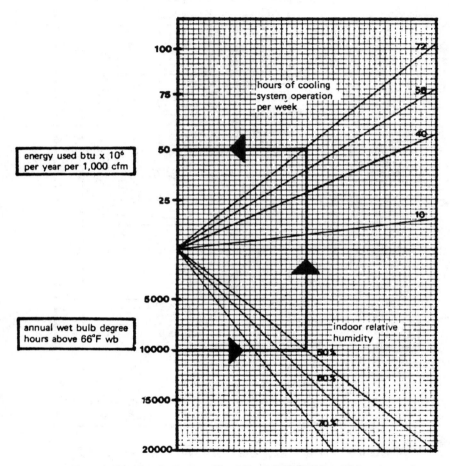

Figure 6-7. Yearly Energy Used Per 1,000 CFM to Maintain
Various Humidity Conditions
(Source: Guidelines For Saving Energy in Existing Buildings--
Building Owners and Operators Manual, ECM-1)

WE degree hours based on 12 Mos/Yr, 8 Hr/Day

Energy used is a function of the WB degree hours above the base of 66F, the RH maintained the No. of hours of controlled humidity. The base RH is 50% which is approximately 78F DB, 66F WB. The figure expresses the energy used per 1000 cfm of air conditioned or dehumidified.

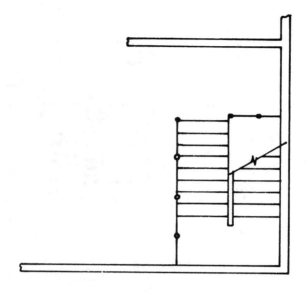

BEFORE

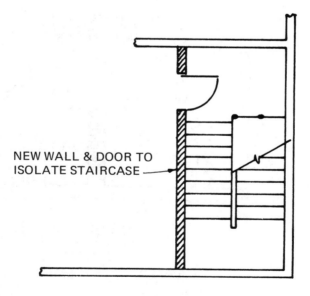

NEW WALL & DOOR TO
ISOLATE STAIRCASE

AFTER

Figure 6-8

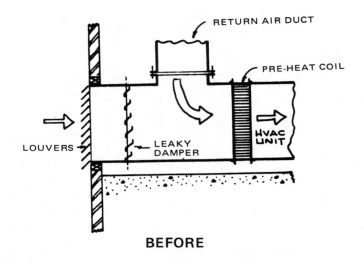

BEFORE

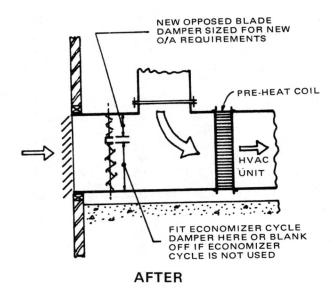

AFTER

Figure 6-9

- Reducing exhaust air by filterizing fitting baffles or a false hood inside existing hood. (See Figure 6-10.)
- Installing a separate make-up air system for hoods. The hood make-up air system would consist of a fan drawing in outdoor air and passing through a heating coil to temper air.

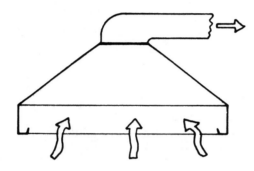

LOW VELOCITY EXHAUST
(HIGH VOLUME)

BEFORE

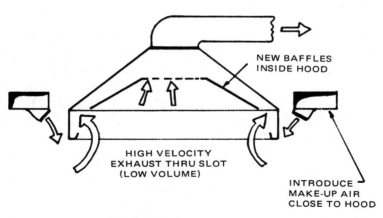

NEW BAFFLES
INSIDE HOOD

HIGH VELOCITY
EXHAUST THRU SLOT
(LOW VOLUME)

INTRODUCE
MAKE-UP AIR
CLOSE TO HOOD

AFTER

Figure 6-10

HEAT FLOW DUE TO CONDUCTION

When a temperature gradient exists on either side of a wall, a flow of heat from hot side to cold side occurs. The flow of heat is defined by Formula 6-3.

$$Q = k/d \cdot A \cdot \Delta T \qquad \qquad Formula\ (6\text{-}3)$$
$$U = k/d = 1/R$$

Where:

Q   is the rate of flow BTUH
d   is the thickness of the material in inches
A   is the area of the wall, ft^2
ΔT is the temperature difference, °F
U   is the conductance of the material—BTU/hr/sq ft/F
k   is the conductivity of the material
R   is the resistance of the material.

Resistance of material in series are additive. Thus the importance of insulation is that it increases the R factor, which in turn reduces the heat flow.

Complete tables for conductors and resistances of various building materials can be found in the **ASHRAE** Guide and Data Book.

HEAT FLOW DUE TO RADIATION

When analyzing a building the conductive portion and radiant portion of heat flow should be treated independently.

Radiation is the transfer of radiant energy from a source to a receiver. The radiation from the source (sun) is partially absorbed by the receiver and partially reflected. The radiation absorbed depends upon its surface emissivity, area, and temperature, as expressed by Formula 6-4.

$$Q = \epsilon\ \sigma\ A\ T^4 \qquad \qquad Formula\ (6\text{-}4)$$

Where:

Q = rate of heat, flow by radiation, BTU/hr
ϵ = emissivity of a body, which is defined as the rate of energy radiated by the actual body. $\epsilon = 1$ for a block body.
σ = Stephen Boltzman's Constant, 1.71×10^{-9} BTU/$ft^2 \cdot$ hr $\cdot T^4$
A = surface area of body in square feet.

In addition the radiant energy causes a greater skin temperature to exist on horizontal surfaces such as the roof. The effect is to cause a greater equivalent ΔT which increases the conductive heat flow. Radiant energy flow through roofs and glass should be investigated since it can significantly increase the heat gain of the building. Radiant energy, on the other hand, reduces HVAC requirements during the heating season.

ENERGY AUDITS OF ROOFS

The handy tables and graphs presented in this section are based on the "sunset" program developed for the ECM-2 Manual. The program was based on internal heat gains of 12 BTU/ square feet/hour when occupied, 10% average outdoor air ventilation when occupied, and one-half air change per hour continuous infiltration. For significantly different conditions an individual computer run should be made using one of the programs listed in Chapter 5.

A summary of heat losses and heat gains for twelve cities is illustrated in Figures 6-11 and 6-12 respectively. The cumulative values shown take into account both conductive and radiant contributions. Thus a dark covered roof will reduce the heat loss during the winter but increases the heat gain in the summer. Usually the cooling load dictates the color of the roof.

To reduce the HVAC load the U-Factor of the roof is increased by adding insulation.

Estimates of savings can be made by using Figures 6-13, 6-14, and 6-15. The figures take into account both radiant effect and the greater ΔT which occurs due to radiant energy. For cooling load considerations the color of the roof is important. Light color roofs, or adding a surface layer of white pebbles or gravel, are sometimes used. (Care should be taken on existing buildings that structural bearing capacity is not exceeded.)

In addition the roof temperature can be lowered by utilizing a roof spray. (Care should be taken that proper drainage and structural considerations are taken into account.)

Solar radiation data is illustrated in Figure 15-1, Chapter 15.

dubin-mindell-bloome-associates
consulting
engineers

YEARLY HEAT LOSS/SQUARE FOOT THROUGH ROOF

| City | Latitude | Solar Radiation Langley's | Degree-Days | Heat Loss Through Roof BTU/Ft2 Year | | | |
| | | | | U=0.19 | | U=0.12 | |
				a=0.3	a=0.8	a=0.3	a=0.8
Minneapolis	45°N	325	8,382	35,250	30,967	21,330	18,642
Denver	40°N	425	6,283	26,794	22,483	16,226	13,496
Concord, N.H.	43°N	300	7,000	32,462	27,678	19,649	16,625
Chicago	42°N	350	6,155	27,489	23,590	16,633	14,190
St. Louis	39°N	375	4,900	20,975	17,438	12,692	10,457
New York	41°N	350	4,871	21,325	17,325	12,911	10,416
San Francisco	38°N	410	3,015	10,551	8,091	6,381	4,784
Atlanta	34°N	390	2,983	12,601	9,841	7,619	5,832
Los Angeles	34°N	470	2,061	4,632	3,696	2,790	2,142
Phoenix	33°N	520	1,765	5,791	4,723	3,487	2,756
Houston	30°N	430	1,600	6,045	4,796	3,616	2,778
Miami	26°N	451	141	259	130	139	55

a is the absorption coefficient of the building material

a = .3 (White)

a = .5 (Light colors such as yellow, green, etc.)

a = .8 (Dark colors)

Figure 6-11. Heat Losses for Roofs

(Source: Guidelines For Saving Energy in Existing Buildings—Engineers, Architects and Operators Manual., ECM-2)

dubin-mindell-bloome-associates
consulting engineers

YEARLY HEAT GAIN/SQUARE FOOT THROUGH ROOF

City	Latitude	Solar Radiation Langley's	D.B. Degree-Hours Above 78°F	Heat Gain Through Roof BTU/Ft² Year			
				U=0.19		U=0.12	
				a=0.3	a=0.8	a=0.3	a=0.8
Minneapolis	45°N	325	2,500	2,008	8,139	1,119	4,728
Concord, N.H.	43°N	300	1,750	1,892	7,379	1,043	4,257
Denver	40°N	425	4,055	2,458	9,859	1,348	5,680
Chicago	42°N	350	3,100	2,104	7,918	1,185	4,620
St. Louis	39°N	375	6,400	4,059	12,075	2,326	7,131
New York	41°N	350	3,000	2,696	9,274	1,534	5,465
San Francisco	38°N	410	3,000	566	5,914	265	3,354
Atlanta	34°N	390	9,400	4,354	14,060	2,482	8,276
Los Angeles	34°N	470	2,000	1,733	10,025	921	5,759
Phoenix	33°N	520	24,448	12,149	24,385	7,258	14,649
Houston	30°N	430	11,500	7,255	20,931	4,176	12,369
Miami	26°N	451	10,771	9,009	24,594	5,315	14,716

a is the absorption coefficient of the building material

a = .3 (White)
a = .5 (Light colors such as yellow, green, etc.)
a = .8 (Dark colors)

Figure 6-12. Heat Gains for Roofs

(Source: Guidelines For Saving Energy in Existing Buildings—Engineers, Architects and Operators Manual, ECM-2)

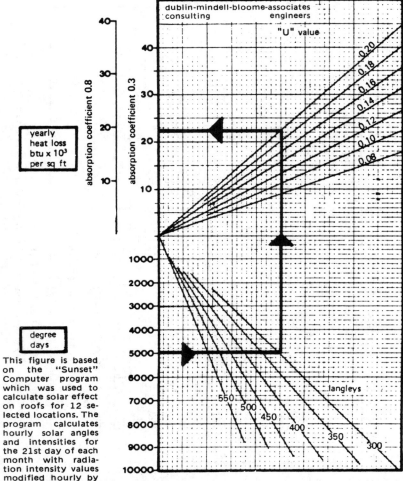

This figure is based on the "Sunset" Computer program which was used to calculate solar effect on roofs for 12 selected locations. The program calculates hourly solar angles and intensities for the 21st day of each month with radiation intensity values modified hourly by the average percentage of cloud cover taken from weather records. Heat losses are based on a 68F indoor temperature.

The solar effect on a roof was calculated using sol-air temperature and the heat entering or leaving a space was calculated using the equivalent temperature difference. Roof mass ranged from 25-35 lbs/sq ft and thermal lag averaged 3½ hours. Additional assumptions were: (1) Total internal heat gain of 12 Btu/sq ft. (2) Outdoor air ventilation rate of 10%. (3) Infiltration rate of 1.2 air change per hour. Daily totals were then summed for the number of days in each month to arrive at monthly heat losses. The length of the heating season for each location considered was determined from weather data and characteristic operating periods. Yearly heat losses were derived by summing monthly totals for the length of the cooling season.

Absorption coefficients and U values were varied and summarized for the 12 locations. The data was then plotted and extrapolated to include the entire range of degree days.

Figure 6-13. Yearly Heat Loss Through Roof
(Source: Guidelines For Saving Energy in Existing Buildings—
Engineers, Architects and Operators Manual, ECM-2)

dubin-mindell-bloome-associates
consulting   engineers

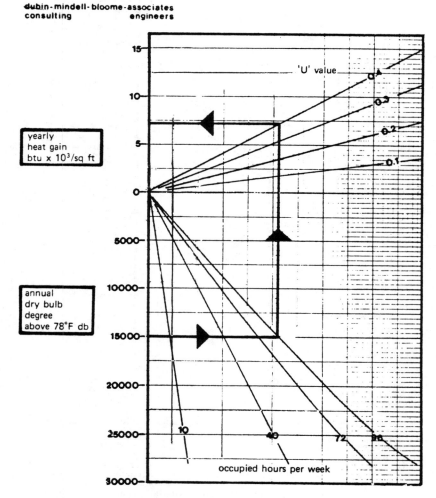

Figure 6-14. Yearly Conduction Heat Gain Through Walls,
Roofs and Floors
(Source: Guidelines For Saving Energy in Existing Buildings—
Engineers, Architects and Operators Manual, ECM-2)

This figure is based on degree hours with a base of 56 hours/week. The figure is based
on the formula: Q (Heat Gain)/yr = Degree Hours/yr X 'U' Value. The major portion
of degree hours occur between 10 a.m. and 3 p.m. Hence for occupancies between
10 and 56 hrs/wk, the degree hour distribution can be assumed to be linear. However,
for occupancies greater than 56 hrs/wk the degree hour distribution becomes nonlinear, particularly in locations with greater than 15,000 degree hours. This is reflected by the curves for 72 and 96 hr/wk occupanices.

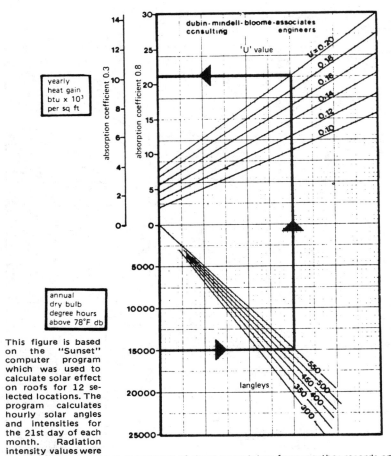

yearly
heat gain
btu x 10³
per sq ft

annual
dry bulb
degree hours
above 78°F db

This figure is based on the "Sunset" computer program which was used to calculate solar effect on roofs for 12 selected locations. The program calculates hourly solar angles and intensities for the 21st day of each month. Radiation intensity values were modified by the average percentage of cloud cover taken from weather records on an hourly basis. Heat gains are based on a 78F indoor temperature.

The solar effect on a roof was calculated using sol-air temperature and the heat entering or leaving a space was calculated with the equivalent temperature difference. Roof mass ranged from 25-35 lbs/sq ft and thermal lag averaged 3½ hours. During the cooling season, internal gains, ventilation, infiltration and conduction through the building skin create a cooling load. The additional load caused by heat gain through the roof was calculated for each day. Daily totals were then summed for the number of days in each month to arrive at monthly heat gains. The length of the cooling season for each location considered was determined from weather data and characteristic operating periods. Yearly heat gains were derived by summing monthly totals for the length of the cooling season.

Absorption coefficients and 'U' values were varied and summarized for the 12 locations. Gains include both the solar and conduction components of heat gain. Values of the conduction heat gain component through roofs were deducted from the total heat gains to derive the solar component. The solar component was then plotted and extrapolated to include the entire range of degree hours.

Figure 6-15. Yearly Solar Heat Gain Through Roof
(Source: Guidelines For Saving Energy in Existing Buildings—
Engineers, Architects and Operators Manual, ECM-2)

COMPUTER SERVICES

Most manufacturers of roof insulation and window treatments offer computer simulations to estimate savings as a result of using their products. These programs are usually available at no cost through authorized distributors and contractors.

For information concerning insulation products and services call the nearest authorized distributor or contact:

> North American Insulation Manufacturers Assoc.
> 48 Canal Center Plaza, Suite 310
> Alexandria, VA 22314

Manufacturers such as 3M offer computer load simulations for their products. For information concerning window treatments contact the manufacturers listed in the next section.

SIM 6-4

An energy audit of the roof indicates the following:

Area	20,000 square feet
Present "R" value =	8
(Estimation based on insulation thickness and type)	
Degree-Days (winter)	3,000
Occupied hrs/week	40
D.B. Degree-Hours above 78°F	9,400
Fuel cost	$5/10^6$ BTU
Boiler efficiency	.65
Electric rate	6¢ per KWH
Air-condition requirement	.8 KW/Ton-Hr
Roof Absorption	.3
Solar radiation	390 Langleys

It is proposed that additional insulation of R=13 should be installed.

Comment on the potential savings.

ANALYSIS

$$R_T = R_1 + R_2 = 8 + 13 = 21$$
$$U_{Before} = 1/8 = .125$$
$$U_{After} = 1/21 = .047$$

Savings

Winter

From Figure 6-13

$Q_{Before} = 7$ BTU $\times 10^3$ per square feet

$Q_{After} = 2$ BTU $\times 10^3$ per square feet

Savings = $(7-2) \times 10^3 \times 20 \times 10^3 \times 5/10^6 / .65 = \$769$

Summer

Conduction

From Figure 6-14

$Q_{Before} = 1 \times 10^3$ BTU/sq ft/yr

$Q_{After} = .4 \times 10^3$ BTU/sq ft/yr

Savings = $(1-.4) \times 10^3 \times 20,000 = 12,000 \times 10^3$ BTU/Yr

Radiation

From Figure 6-15

$Q_{Before} = 8.5 \times 10^3$ BTU/sq ft/yr

$Q_{After} = 2.5 \times 10^3$ BTU/sq ft/yr

Savings = $(8.5-2.5) \times 10^3$ BTU/sq ft/yr $\times 20,000 =$
$\qquad 120,000 \times 10^3$ BTU/yr

$$\text{Savings} = \frac{(120,000 + 12,000)}{12,000} \times 10^3 \times .8 \times .06 = \$528/yr$$

Total Savings = $1297/year.

Figure 6-16 illustrates typical insulation conductance values recommended based on degree-day data.

INSULATION VALUE FOR HEAT FLOW THROUGH OPAQUE AREAS OF ROOFS AND CEILINGS

Heating Season Degree-Days	U value (Btu/hr/sq ft/$^{\circ}$F)
1 - 1000	0.12
1001 - 2000	0.08
2001 and above	0.05

INSULATION VALUE FOR HEAT FLOW THROUGH OPAQUE EXTERIOR WALLS FOR HEATED AREAS

Heating Seasons Degree-Days	U value (Btu/hr/sq ft/$^{\circ}$F)
0 - 1000	0.30
1001 - 2500	0.25
2500 - 5000	0.20
5000 - 8000	0.15

Cooling Season — The recommended U value of insulation for heat flow through exterior roofs, ceilings, and walls should be less than 0.15 Btu/hr/sq ft/ $^{\circ}$F.

Figure 6-16. Insulation Conductance Values for Roofs and Walls
(Source: Instructions For Energy Auditors)

THE GLASS AUDIT

CONDUCTION CONSIDERATIONS

Glass traditionally has poor conductance qualities and accounts for significant heat gains due to radiant energy.

To estimate savings as a result of changing glass types, Figures 6-17 through 6-22 can be used. Figures 6-19 and 6-20 illustrate the heat loss and gain due to conduction for winter and summer respectively. Figures 6-21 and 6-22 can be used to calculate the radiant heat gain during summer.

To decrease losses due to conductance either the glass needs to be replaced, modified, or an external thermal blanket added. Descriptions of window treatments are discussed at the end of the chapter.

dubin-mindell-bloome-associates
consulting   engineers

| City | Latitude | Solar Radiation Langleys | Degree-Days | Heat Loss Through Window BTU/ft² Year | | | | | |
| | | | | North | | East & West | | South | |
				Single	Double	Single	Double	Single	Double
Minneapolis	45°N	325	8,382	187,362	94,419	161,707	84,936	140,428	74,865
Concord, N.H.	43°N	300	7,000	158,770	83,861	136,073	73,303	122,144	67,586
Denver	40°N	425	6,283	136,452	70,449	117,487	62,437	109,365	59,481
Chicago	42°N	350	6,155	147,252	75,196	126,838	65,810	110,035	58,632
St. Louis	39°N	375	4,900	109,915	56,054	94,205	49,355	84,399	45,398
New York	41°N	350	4,871	109,672	54,986	93,700	48,611	82,769	44,580
San Francisco	38°N	410	3,015	49,600	25,649	43,866	23,704	41,691	23,239
Atlanta	34°N	390	2,983	63,509	31,992	55,155	28,801	51,837	28,092
Los Angeles	34°N	470	2,061	21,059	11,532	19,487	10,954	19,485	10,989
Phoenix	33°N	520	1,765	25,951	14,381	22,381	12,885	22,488	12,810
Houston	30°N	430	1,600	33,599	17,939	30,744	17,053	30,200	16,861
Miami	26°N	451	141	1,404	742	1,345	742	1,345	742

Figure 6-17. Yearly Heat Loss/Square Foot of Single Glazing and Double Glazing
(Source: Guideline For Saving Energy in Existing Buildings—Engineers, Architects and Operators Manual, ECM-2)

dubin-mindell-bloome-associates
consulting engineers

City	Latitude	Solar Radiation Langleys	D.B. Degree-Hours Above 78°F	Heat Gain Through Window BTU/ft² Year					
				North		East & West		South	
				Single	Double	Single	Double	Single	Double
Minneapolis	45°N	325	2,500	36,579	33,089	98,158	88,200	82,597	70,729
Concord, N.H.	43°N	300	1,750	33,481	30,080	91,684	82,263	88,609	76,517
Denver	40°N	425	4,055	44,764	39,762	122,038	108,918	100,594	85,571
Chicago	42°N	350	3,100	35,595	31,303	93,692	83,199	87,017	74,497
St. Louis	39°N	375	6,400	55,242	45,648	130,018	112,368	103,606	85,221
New York	41°N	350	3,000	40,883	35,645	109,750	97,253	118,454	102,435
San Francisco	38°N	410	3,000	29,373	28,375	88,699	81,514	73,087	64,169
Atlanta	34°N	390	9,400	59,559	50,580	147,654	129,391	106,163	87,991
Los Angeles	34°N	470	2,000	47,912	43,264	126,055	112,869	112,234	97,284
Phoenix	33°N	520	24,448	137,771	97,565	242,586	191,040	211,603	131,558
Houston	30°N	430	11,500	88,334	72,474	213,739	184,459	188,718	156,842
Miami	26°N	451	10,771	98,496	79,392	237,763	203,356	215,382	179,376

Figure 6-18. Yearly Heat Gain/Square Foot of Single Glazing and Double Glazing
(Source: Guidelines For Saving Energy in Existing Buildings—Engineers, Architects and Operators Manual, ECM-2)

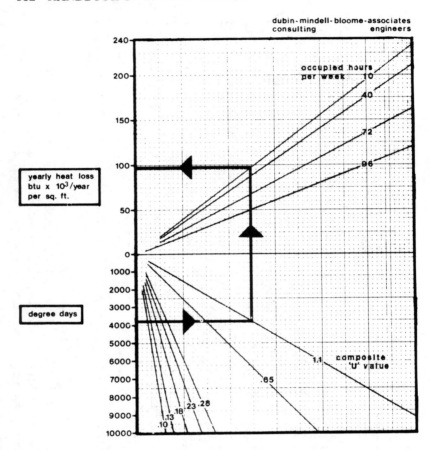

NOTE

The development of this figure was based on the assumptions that

1. Thermal barriers are closed only when the building is unoccupied
2. The average degree-day distribution is 25% during the daytime and 75% during nighttime.

The number of degree days occurring when the thermal barriers are closed (adjusted degree-days — DD_A) were determined from the characteristic occupancy periods shown in the figure. This can be expressed as a fraction of the total degree-days (DD_T) by the relationship:

$$DD_A = 0.25\ DD_T \left(\frac{\text{unoccupied daytime hours/week}}{\text{total daytime hours/week}} \right) + 0.75\ DD_T \left(\frac{\text{unoccupied nighttime hours/week}}{\text{total nighttime hours/week}} \right)$$

Yearly heat losses can then be determined by: Q (heat loss/yr) = DD_A X U value X 24

Figure 6-19. Yearly Heat Loss for Windows with Thermal Barriers
(Source: Guidelines For Saving Energy in Existing Buildings—Engineers, Architects and Operators Manual, ECM-2)

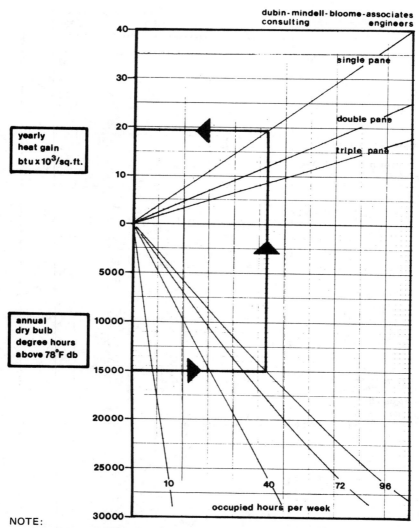

NOTE:

The figure is based on the formula: Q (heat gain)/yr = degree hours/yr X U value. U values assumed were 1.1 for single pane, 0.65 for double pane and 0.47 for triple pane. The major portion of degree hours occur between 10 a.m. and 3 p.m. Hence, for occupancies between 10 and 56 hours/week, the degree hour distribution can be assumed to be linear. However, for occupancies greater than 56 hours/week the degree hour distribution becomes nonlinear, particularly in locations with greater than 15,000 degree hours. This is reflected by the curves for 72 and 96 hour/week occupancies.

Figure 6-20. Yearly Conduction Heat Gain Through Windows
(Source: Guidelines For Saving Energy in Existing Buildings—Engineers, Architects and Operators Manual, ECM-2)

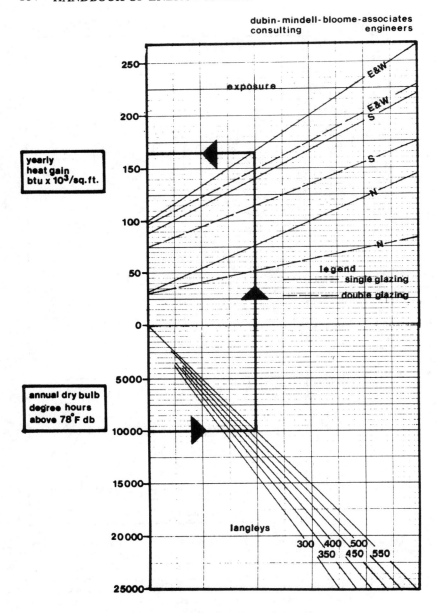

(See page 152 for notes on this figure)

Figure 6-21. Yearly Solar Heat Gain Through Windows
Latitude 25°N – 35°N

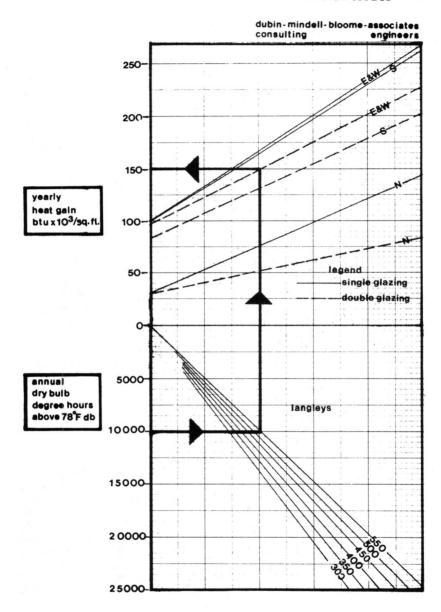

(See page 152 for notes on this figure)

**Figure 6-22. Yearly Solar Heat Gain Through Windows
Latitude 35°N — 45°N**

SOLAR RADIATION CONSIDERATIONS

In addition to heat flow due to conduction, a significant heat flow occurs through glass due to the sun's radiant energy. The radiant energy will decrease heating requirements during the winter time but greatly increase the air-conditioning load during the cooling season.

To reduce solar loads, several common devices are used.
- Roller shades (least expensive)
- Reflective polyester film
- Venetion blinds
- Vertical louver blinds
- External louvered screens
- Tinted or reflective glass (most expensive)

Descriptions of window treatments are discussed at the end of this chapter.

To determine the energy saved from shading devices, Figures 6-21 and 6-22 can be used. Occupancy for these figures is based on 5 days/week, 12 hours/day. If space is occupied differently, prorate the results. The savings for window treatments is estimated by multiplying the annual heat gain by the shading coefficient of the window treatment.

NOTES FOR FIGURES 6-21 AND 6-22

These figures are based on the "Sunset" Computer program which was used to calculate solar effect on windows for 12 locations. The program calculates hourly solar angles and intensities for the 21st day of each month. Radiation intensity values were modified by the average percentage of cloud cover taken from weather records on an hourly basis. Heat gains are based on a 78°F indoor temperature. During the cooling season, internal gains, ventilation, infiltration and conduction through the building can create a cooling load. The additional load caused by heat gain through the windows was calculated for each day. Daily totals were then summed for the number of days in each month to arrive at monthly heat gains. The length of the cooling season for each location considered was determined from weather data and characteristic operating periods. Yearly heat gains were derived by summing monthly totals for the length of the cooling season. Gains are based purely on the solar component. The solar component was then plotted and extrapolated to include the entire range of degree hours. The heat gains assume that the windows are subjected to direct sunshine. If shaded, gains should be read from the north exposure line. The accuracy of the graph diminishes for location with less than 5,000 degree hours.

WINDOW TREATMENTS

Several types of window treatments to reduce losses have become available. This section describes some of the products on the market based on information supplied by manufacturers. No claims are made concerning the validity or completeness described. The summary is based on "Windows For Energy Efficient Buildings" as prepared by the Lawrence Berkeley Laboratory for U.S. DOE.

Window treatment manufacturers offer a wide variety of "free programs" to help in evaluation of their products.

Solar Control

Solar Control Films—A range of tinted and reflective polyester films are available to adhere to inner window surfaces to provide solar control for existing clear glazing. Films are typically two- or three-layer laminates composed of metalized, transparent and/or tinted layers. Films are available with a wide range of solar and visible light transmittance values, resulting in shading coefficients as low as 0.24. Most films are adhered with precoated pressure sensitive adhesives. Reflective films will reduce winter U values by about 20%. (Note that a new solar control film, which provides a U value of 0.68, is described in the Thermal Barriers section below. Films adhered to glass improve the shatter resistance of glazing and reduce transmission, thus reducing fading of furnishings.

Fiber Glass Solar Control Screens—Solar control screen provides sun and glare control as well as some reduction in winter heat loss. Screens are woven from vinyl-coated glass strands and are available in a variety of colors. Depending on color and weave, shading coefficients of 0.3-0.5 are achieved. Screens are durable, maintenance free, and provide impact resistance. They are usually applied on the exterior of windows and may (1) be attached to mounting rails and stretched over windows, (2) mounted in rigid frames and installed over windows, or (3) made into roller shades which can be retracted and stored as

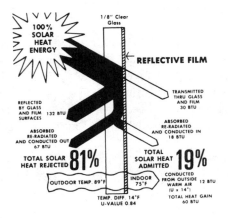

Figure 6-23. Application of Window Film
(Photograph courtesy of 3M Company)

desired. Names of local distributors, installers, and retailers may be obtained by writing to major fabric manufacturers.

Motorized Window Shading System— A variety of plastic and fabric shades is available for use with a motorized window shading system. Reversible motor is located within the shade tube roller and contains a brake mechanism to stop and hold in any position. Motor controls may be gauged and operated locally or from a master station. Automatic photoelectric controls are available that (1) monitor sun intensity and angle and adjust shade position to provide solar control and (2) employ an internal light sensor and provide a preset level of internal ambient light.

Exterior Sun Control Louvers—Operable external horizontal and vertical louver systems are offered for a variety of building sun control applications. Louvers are hinged together and can be rotated in unison to provide the desired degree of shading for any sun position. Operation may be manual or electric; electrical operation may be initiated by manual switches, time clock, or sun sensors. Louvers may be closed to reduce night thermal losses. Sun control elements are available in several basic shapes and in a wide range of sizes.

External Venetian Blinds—Externally mounted, all weather, venetian blinds may be manually operated from within a building or electrically operated and controlled by means of automatic sun sensors, time controls, manual switches, etc. Aluminum slats are held in position with side guides and controlled by weatherproof lifting tapes. Slats can be tilted to modulate solar gain, closed completely, or restricted to admit full light and heat. Blinds have been in use in Europe for many years and have been tested for resistance to storms and high winds.

Adjustable Louvered Windows—Windows incorporating adjustable external louvered shading devices. Louvers are extruded aluminum or redwood, 3 to 5 inches wide, and are manually controlled. Louvers may be specified on double-hung, hinged,

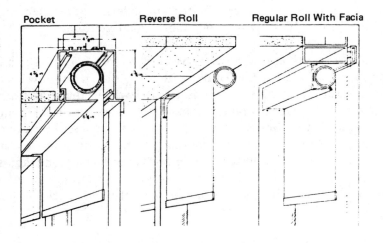

Figure 6-24. Motorized Shade System

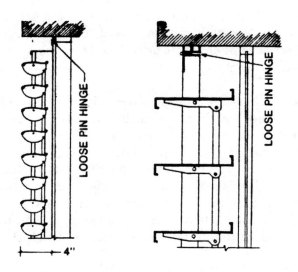

Figure 6-25. Mounting of External Louvers

Figure 6-26. Installation of Louvered Solar Screens

Greenwich Harbor, Greenwich, Connecticut is seen through a louvered solar screen at left and through 34 conventional venetian blind louvers on the right. The two panels are otherwise identical, having the same slat angle and ratio of louver width to louver spacing.

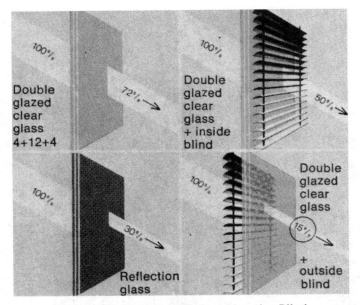

Figure 6-27. Exterior "L"-Type Venetian Blind

or louvered-glass windows. When open, the louvers provide control of solar gain and glare; when closed, they provide privacy and security.

Solar Shutters—The shutter is composed of an array of aluminum slats set at 45° or 22½° from the vertical to block direct sunlight. Shutters are designed for external application and may be mounted vertically in front of window or projected outward from the bottom of the window. Other rolling and hinged shutters are stored beside the window and roll or swing into place for sun control, privacy, or security.

Thermal Barriers

Multilayer, Roll-Up Insulating Window Shade—A multilayer window shade which stores in compact roll and utilizes spacers to separate the aluminized plastic layers in the deployed position, thereby creating a series of dead air spaces. Five-layer shade combined with insulated glass provides R8 thermal resistance. Figure 6-28 illustrates a thermal window shade.

Figure 6-28. Thermal Window Shade Installation

Insulating Window Shade—ThermoShade thermal barrier is a roll-up shade composed of hollow, lens-shaped, rigid white PVC slats with virtually no air leakage through connecting joints. Side tracking system reduces window infiltration. Designed for interior installation and manual or automatic operation. When added to a window, the roll-up insulating shade provides R4.5 for single-glazed window or R5.5 for double-glazed window. Quilt is composed of fabric outer surfaces and two polyester fiberfill layers sandwiched around a reflective vapor barrier. Quilt layers are ultrasonically welded. Shade edges are enclosed in a side track to reduce infiltration.

Reflective, Perforated Solar Control Laminate—Laminate of metalized weatherable polyester film and black vinyl which is then perforated with 225 holes/in^2, providing 36% open area. Available in a variety of metallized and nonmetallized colors, the shading coefficients vary from 0.30 to 0.35 for exter-

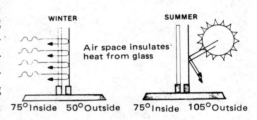

WINTER   SUMMER

Air space insulates heat from glass

75°Inside 50°Outside   75°Inside 105°Outside

nally mounted screens and 0.37 to 0.45 for the material adhered to the inner glass surface. The laminate is typically mounted in aluminum screen frames which are hung externally, several inches from the window; it can also be utilized in a roll-up form. Some reduction in winter U value can be expected with external applications.

Semi-Transparent Window Shades—Roll-up window shades made from a variety of tinted or reflective solar control film laminates. These shades provide most of the benefits of solar control film applied directly to glass but provide additional flexibility and may be retracted on overcast days or when solar gain is desired. Shades available with spring operated and gravity (cord and reel) operated rollers as well as motorized options.

Shading coefficients as low as 0.13 are achieved and a tight fitting shade provides an additional air space and thus reduced U-value.

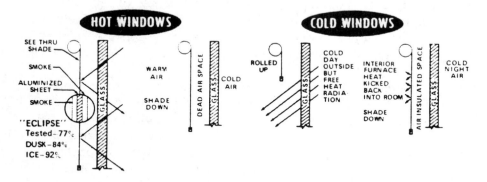

Louvered Metal Solar Screens—Solar screen consists of an array of tiny louvers which are formed from a sheet of thin aluminum. The louvered aluminum sheet is then installed in conventional screen frames and may be mounted against a window in place of a regular insect screen or mounted away from the building to provide free air circulation around the window. View to the outside is maintained while substantially reducing solar gain. Available in a light green or black finish with shading coefficients of 0.21 or 0.15, respectively.

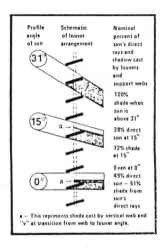

Operable External Louver Blinds—Solar control louver blinds, mounted on the building exterior, can be controlled manually or automatically by sun and wind sensors. Slats can be tilted to modulate light, closed completely, or retracted to admit full light and heat. Developed and used extensively in Europe, they provide summer sun control, control of natural light, and reduction of winter heat loss.

Louvered Metal Solar Screens—Solar screen consists of an array of tiny fixed horizontal louvers which are woven in place. Louvers are tilted at 17° to provide sun control. Screen material is set in metal frames which may be permanently installed in a variety of configurations or designed for removal. Installed screens have considerable wind and impact resistance. Standard product (17 louvers/inch) has a shading coefficient of 0.23; low sun angle variant (23 louvers/inch) has a shading coefficient of 0.15. Modest reductions in winter U value have been measured.

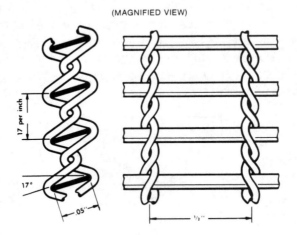

(MAGNIFIED VIEW)

A comparison of visibility with the louvered screens against conventional venetian blinds is illustrated in Figure 6-26.

Insulating Solar Control Film—A modified solar control film designed to be adhered to the interior of windows provides conventional solar control function and has greatly improved insulating properties. Film emissivity is 0.23-0.25 resulting in a U-value of 0.68 Btu/ft² hr-°F under winter conditions, compared to 0.87 for conventional solar control films and 1.1 for typical single-glazed windows.

Interior Storm Window— Low cost, do-it-yourself interior storm window with a rigid plastic glazing panel. Glazing panel may be removed for cleaning or summer storage. Reduces infiltration losses as well as conductive/convective heat transfer.

Retrofit Insulating Glass System—Single glazing is converted to double glazing by attaching an extra pane of glass with neoprene sealant. A dessicant-filled aluminum spacer absorbs moisture between the panes. An electric resistance wire embedded in the neoprene is heated with a special power source. This hermetically seals the window. New molding can then be applied if desired.

Infiltration

Weather-Strip Tape—A polypropylene film scored along its centerline so that it can be easily formed into a "V" shape. It has a pressure sensitive adhesive on one leg of the "V" for application to seal cracks around doors and windows. On an average fitting, double-hung window it will reduce infiltration by over 70%. It can be applied to rough or smooth surfaces.

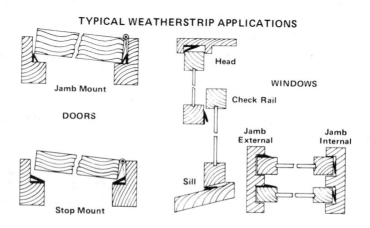

TYPICAL WEATHERSTRIP APPLICATIONS

PASSIVE SOLAR BUILDING DESIGNS

A passive solar system is defined as one in which thermal energy flows by natural means. Examples of solar building design include:

- Solar greenhouses which are built on the south side of buildings. These can produce 60-100% of heating and cooling requirements.
- Underground buildings which use ground temperature to provide year-round temperature requirements.
- Enhanced natural ventilation through solar chimneys or use of "Trombe wall."

In these examples and others passive systems accomplish work (heating and cooling) by natural means such as gravity flows, thermosiphons, etc.

To study how the building reacts to loads, its storage effect, etc. computer simulations are many times used. One such system is described below:

PEGFIX—predicts auxiliary heat demand and excess heat available in a space with user-defined maximum and minimum air temperatures. The program is directly useful in sizing and specifying system components and auxiliary equipment. Results stored by PEGFIX are: total auxiliary heating load, excess heat available, maximum fan rate required to remove excess heat, and maximum hourly auxiliary load.

PEGFLOAT—predicts hourly temperatures of air and storage mass in a space without auxiliary heat input or removal of excess heat. Its purpose is to evaluate temperature excursions in a 100% solar-dependent operating mode. This program can examine non-south glazing orientations with user-specified hourly values for insolation. PEGFLOAT automatically stores maximum and minimum air and storage temperatures of the system modelled.

Both programs require few user-defined inputs regarding the building design and local weather: heat loss coefficients; effective thermal capacity and storage surface area; solar energy available, fraction to storage and fraction to air; average outdoor temperature and daily range. Programs differentiate day and night heat loss values, and can automatically proportion daylong insolation. Each can be run through a 24-hour day, without user interaction, in five to nine minutes. Hourly values of air and storage temperatures, and auxiliary or excess heat, can be displayed without interrupting program execution. Optional hourly display does not affect data storage.

REDUCING STRATIFIED AIR

As indicated in this chapter both the HVAC and building envelope considerations must be considered. An example of this system approach occurs when heat stratification near ceilings is reduced.

One way of reducing air temperatures near ceilings during the heating season is to use a circulation fan with connected ductwork, as illustrated in Figure 6-29.

The result of reducing ceiling temperature is a reduction in conduction and exhaust losses.

SIM 6-5

Comment on reducing the stratified air temperature from 90°F to 75°F during the heating season.

U = .1

Area = 20,000 square feet

Assume an outside temperature of 15°F and exhaust CFM of 20,000.

BRING YOUR HEAT FROM HERE...

ANALYSIS

A handy formula to relate heat loss from CFM exhausts is:

$$Q = 1.08 \text{ BTU Min./Hr, Ft}^3, \text{F} \times \Delta T$$

Before change

$$Q_{conduction} = U\, A\Delta T = .1 \times 20,000 \times 75 = 150,000 \text{ BTU/H}$$

$$Q_{CFM} = 1.08 \times 20,000 \times 75 = 1,620,000 \text{ BTU/H}$$

$$Q_T = 1,770,000 \text{ BTU/H}$$

After change in stratification temperature

$$Q_c = .1 \times 20,000 \times 60 = 120,000 \text{ BTU/H}$$

$$Q_{CFM} = 1.08 \times 20,000 \times 60 = 1,296,000 \text{ BTU/H}$$

$$Q_T = 1,416,000 \text{ BTU/H}$$

% savings — heating season
= 100 − 1,416,000/1,770,000
   × 100 = 20%

Figure 6-29.
Reducing Air Stratification
Temperatures
(Photograph courtesy of Rusth Industries, Beaverton, Oregon)

7

The Electrical System Energy Audit

The Electrical System Energy Audit (ESEA) requires gathering the following data:
1. Electrical Rate Tariff
2. Existing Lighting System Survey
3. Distribution System Characteristics
    a. Motor Loads
    b. Power Factor and Demand

Typical information gathering forms are illustrated in Chapter 15, Figures 15-15 through 15-18.

Before an ESEA can be analyzed, a thorough knowledge of how electricity is billed needs to be known. The ESEA should uncover what items should be implemented and the effective cost benefit analysis.

Each component of the ESEA is discussed in this section.

ELECTRICAL RATE TARIFF

The basic electrical rate charges contain the following elements:

Billing Demand—The maximum kilowatt requirement over a 15-, 30-, or 60-minute interval.

Load Factor—The ratio of the average load over a designated period to the peak demand load occurring in that period.

Power Factor—The ratio of resistive power to apparent power. Traditionally electrical rate tariffs have a decreasing kilowatt

181

hour (KWH) charge with usage. This practice is likely to gradually phase out. New tariffs are containing the following elements:

Time of Day—Discounts are allowed for electrical usage during off-peak hours.

Ratchet Rate—The billing demand is based on 80-90% of peak demand for any one month. The billing demand will remain at that ratchet for 12 months even though the actual demand for the succeeding months may be less.

The effect of changes in electrical rate tariffs can be significant as illustrated in SIM 7-1.

SIM 7-1

The existing rate structure is as follows:

Demand Charge:

First	25 KW of billing demand	$4.00 per KW per month
Next	475 KW of billing demand	$3.50 per KW per month
Next 1000 KW		$3.25 per KW per month

Energy Charge:

First	2,000 KW-Hrs per month	8¢ per KWH
Next	18,000 KW-Hrs per month	6¢ per KWH
Next 180,000 KW-Hrs per month		4.4¢ per KWH
Etc.		

The new proposed schedule deletes price breaks for usage.

	Billing Months June—September	Billing Months October—May
Demand Charge	$13.00 per KW/Month	$5.00 per KW/Month
Energy Charge	5¢ per KWH	3¢ per KWH

Demand charge based on greatest billing demand month.

Comment on the proposed billing as it would affect an industrial customer who uses 475 kW per hour for 330 hours per month. For the 8 months of winter, demand is 900 kW. For the 4 months of summer, demand is 1200 kW.

ANALYSIS

The proposed rate schedule has two major changes. First, billing demand is on a ratchet basis and discourages peak demand during summer months. The high demand charge encourages the plant to improve the overall load factor. The increased demand charge is partially offset with a lower energy usage rate.

Original Billing

Winter: First 25 KW	$ 100	
Demand Next 475 KW	1,660	
Next 400 KW	1,300	
	$3,060	

Summer: $4,035

Total Demand: 8 X 3060 +4 X 4035 = $40,620

Usage Charge: 475 KW X 330 Hours = 156,750 KWH

First	2,000 KWH @ 8¢ =	$ 160
Next	18,000 KWH @ 6¢ =	1,080
Next 136,750 KWH @ 4.4¢ =		6,016
		$7,256

Total Usage: 7256 X 12 = $87,072

Total Charge = $127,692 or 8.1¢ per KWH

Proposed Billing

Demand: 1200 X $13.00 X 4 months =	$ 62,000	
1200 X $ 5.00 X 8 months =	$ 48,000	
Total Demand:	$110,400	
Usage: 475 KW X 330 X 5¢ X 4 =	$ 31,350	
475 KW X 330 X 3¢ X 8 =	$ 37,620	
Total Usage:	$ 68,970	

Total Charge = $179,370 or 11.4¢ per KWH or a 40% increase

LIGHTING SYSTEM AUDIT

To perform a lighting audit the following information is required:

Room Classification—office, storage, etc.

Room Characteristics—height, width, length, and color

Fixture Characteristics—type of lamp, fixture mounting height, ballast and lamp wattage, plus measured footcandle level.

This information will not only give crude measurements of performance such as watts/square foot, but will also provide sufficient data for a technical analysis. The most common method used to calculate illumination requirements is referred to as the lumen method.

LIGHTING EFFICIENCY

Lighting Basics

About 20 percent of all electricity generated in the United States today is used for lighting.

By understanding the basics of lighting design, several ways to improve the efficiency of lighting systems will become apparent.

There are two common lighting methods used. One is called the "Lumen" method, while the other is the "Point by Point" method. The Lumen method assumes an equal footcandle level throughout the area. This method is used frequently by lighting designers since it is simplest; however, it wastes energy, since it is the light "at the task" which must be maintained and not the light in the surrounding areas. The "Point by Point" method calculates the lighting requirements for the task in question.

The "Point by Point" method makes use of the inverse-square law, which states that the illuminance at a point on a surface perpendicular to the light ray is equal to the luminous intensity of the source at that point divided by the square of the distance between the source and the point of calculation, as illustrated in Formula 7-1.

$$E = \frac{I}{D^2} \qquad \qquad Formula\ (7\text{-}1)$$

Where
- E = Illuminance in footcandles
- I = Luminous intensity in candles
- D = Distance in feet between the source and the point of calculation.

If the surface is not perpendicular to the light ray, the appropriate trigonometrical functions must be applied to account for the deviation.

Lumen Method

A footcandle is the illuminance on a surface of one square foot in area having a uniformly distributed flux of one lumen. From this definition, the "Lumen Method" is developed and illustrated by Formula 7-2.

$$N = \frac{F_1 \times A}{Lu \times L_1 \times L_2 \times Cu} \qquad \textit{Formula (7-2)}$$

Where
- N is the number of lamps required.
- F_1 is the required footcandle level at the task. A footcandle is a measure of illumination; one standard candle power measured one foot away.
- A is the area of the room in square feet.
- Lu is the Lumen output per lamp. A Lumen is a measure of lamp intensity: its value is found in the manufacturer's catalogue.
- Cu is the coefficient of utilization. It represents the ratio of the Lumens reaching the working plane to the total Lumens generated by the lamp. The coefficient of utilization makes allowances for light absorbed or reflected by walls, ceilings, and the fixture itself. Its values are found in the manufacturer's catalogue.
- L_1 is the lamp depreciation factor. It takes into account that the lamp Lumen depreciates with time. Its value is found in the manufacturer's catalogue.
- L_2 is the luminaire (fixture) dirt depreciation factor. It

takes into account the effect of dirt on a luminaire and varies with type of luminaire and the atmosphere in which it is operated.

The Lumen method formula illustrates several ways lighting efficiency can be improved.

Faced with the desire to reduce their energy use,[1] lighting consumers have four options: i) reduce light levels, ii) purchase more efficient equipment, iii) provide light when needed at the task at the required level, and iv) add control and reduce lighting loads automatically. The multitude of equipment options to meet one or more of the above needs permits the consumer and the lighting designer-engineer to consider the trade-offs between the initial and operating costs based upon product performance (life, efficacy, color, glare, and color rendering).

Some definitions and terms used in the field of lighting will be presented to help consumers evaluate and select lighting products best suiting their needs. Then, some state-of-the-art advances will be characterized so that their benefits and limitations are explicit.

LIGHTING TERMINOLOGY

Efficacy — Is the amount of visible light (lumens) produced for the amount of power (watts) expended. It is a measure of the efficiency of a process but is a term used in place of efficiency when the input (W) has different units than the output (lm) and expressed in lm/W.

Color Temperature — A measure of the color of a light source relative to a black body at a particular temperature expressed in degrees Kelvin (°K). Incandescents have a low color temperature (~2800°K) and have a red-yellowish tone; daylight has a high color temperature (~6000°K) and appears bluish. Today, the phosphors used in fluorescent lamps can be blended to provide any desired color temperature in the range from 2800°K to 6000°K.

Color Rendering — A parameter that describes how a light source renders a set of colored surfaces with respect to a black

[1] Source: *Lighting Systems Research*, R. R. Verderber.

body light source at the same color temperature. The color rendering index (CRI) runs from 0 to 100. It depends upon the specific wavelengths of which the light is composed. A black body has a continuous spectrum and contains all of the colors in the visible spectrum. Fluorescent lamps and high intensity discharge lamps (HID) have a spectrum rich in certain colors and devoid in others. For example, a light source that is rich in blues and low in reds could appear white, but when reflected from a substance, it would make red materials appear faded. The same material would appear different when viewed with an incandescent lamp, which has a spectrum that is rich in red.

LIGHT SOURCES[2]

Figure 7-1 indicates the general lamp efficiency ranges for the generic families of lamps most commonly used for both general and supplementary lighting systems. Each of these sources is discussed briefly here. It is important to realize that in the case of fluorescent and high intensity discharge lamps, the figures quoted for "lamp efficacy" are for the lamp only and do not include the associated ballast losses. To obtain the total system efficiency, ballast input watts must be used rather than lamp watts to obtain an overall system lumen per watt figure. This will be discussed in more detail in a later section.

Incandescent lamps have the lowest lamp efficacies of the commonly used lamps. This would lead to the accepted conclusion that incandescent lamps should generally not be used for large area, general lighting systems where a more efficient source could serve satisfactorily. However, this does not mean that incandescent lamps should never be used. There are many applications where the size, convenience, easy control, color rendering, and relatively low cost of incandescent lamps are suitable for a specific application.

General service incandescent lamps do not have good lumen maintenance throughout their lifetime. This is the result of the tungsten's evaporation off the filament during heating as it

---

[2] Source: *Selection Criteria for Lighting Energy Management*, Roger L. Knott.

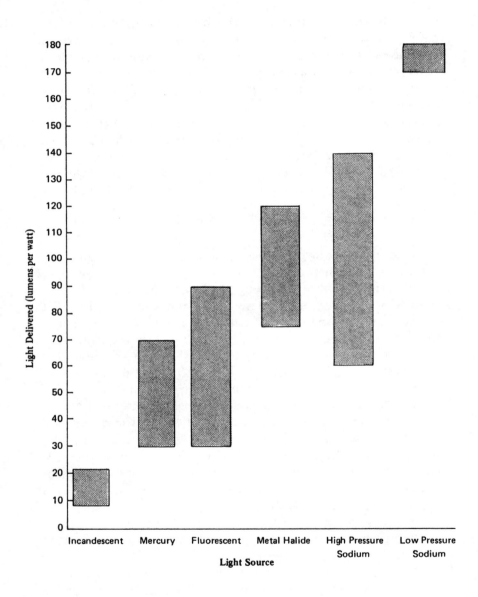

Figure 7-1. Efficiency of Various Light Sources

deposits on the bulb wall, thus darkening the bulb and reducing the lamp lumen output.

Efficient Types of Incandescents for Limited Use

Attempts to increase the efficiency of incandescent lighting while maintaining good color rendition have led to the manufacture of a number of energy-saving incandescent lamps for limited residential use.

Tungsten Halogen—These lamps vary from the standard incandescent by the addition of halogen gases to the bulb. Halogen gases keep the glass bulb from darkening by preventing the filament's evaporation, thereby increasing lifetime up to four times that of a standard bulb. The lumen-per-watt rating is approximately the same for both types of incandescents, but tungsten halogen lamps average 94% efficiency throughout their extended lifetime, offering significant energy and operating cost savings. However, tungsten halogen lamps require special fixtures, and during operation the surface of the bulb reaches very high temperatures, so they are not commonly used in the home.

Reflector or R-Lamps—Reflector lamps are incandescents with an interior coating of aluminum that directs the light to the front of the bulb. Certain incandescent light fixtures, such as recessed or directional fixtures, trap light inside. Reflector lamps project a cone of light out of the fixture and into the room, so that more light is delivered where it is needed. In these fixtures, a 50-watt reflector bulb will provide better lighting and use less energy when substituted for a 100-watt standard incandescent bulb.

Reflector lamps are an appropriate choice for task lighting (because they directly illuminate a work area) and for accent lighting. Reflector lamps are available in 25, 30, 50, 75, and 150 watts. While they have a lower initial efficiency (lumens per watt) than regular incandescents, they direct light more effectively, so that more light is actually delivered than with regular incandescents. (See Figure 7-2.)

Standard Incandescent

R-Lamp

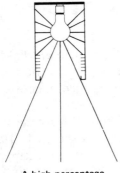

A high percentage
of light output
is trapped in fixture

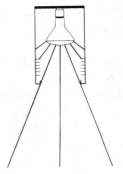

An aluminum
coating directs light
out of the fixture

ER Lamp

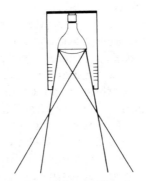

The beam is focused 2 inches
ahead of the lamp, so that very
little light is trapped in the fixture

Figure 7-2. Comparison of Incandescent Lamps

PAR Lamps—Parabolic aluminized reflector (PAR) lamps are reflector lamps with a lens of heavy, durable glass, which makes them an appropriate choice for outdoor flood and spot lighting. They are available in 75, 150, and 250 watts. They have longer lifetimes with less depreciation than standard incandescents.

ER Lamps—Ellipsoidal reflector (ER) lamps are ideally suited for recessed fixtures, because the beam of light produced is focused two inches ahead of the lamp to reduce the amount of light trapped in the fixture. In a directional fixture, a 75-watt ellipsoidal reflector lamp delivers more light than a 150-watt R-lamp. (See Figure 7-2.)

Mercury vapor lamps find limited use in today's lighting systems because fluorescent and other high intensity discharge (HID) sources have surpassed them in both lamp efficacy and system efficiency. Typical ratings for mercury vapor lamps range from about 30 to 70 lumens per watt. The primary advantages of mercury lamps are a good range of color, availability, in sizes as low as 30 watts, long life and relatively low cost. However, fluorescent systems are available today which can do many of the jobs mercury used to do and they do it more efficiently. There are still places for mercury vapor lamps in lighting design, but they are becoming fewer as technology advances in fluorescent and higher efficacy HID sources.

Fluorescent lamps have made dramatic advances in the last 10 years. From the introduction of reduced wattage lamps in the mid-1970s, to the marketing of several styles of low wattage, compact lamps recently, there has been a steady parade of new products. Lamp efficacy now ranges from about 30 lumens per watt to near 90 lumens per watt. The range of colors is more complete than mercury vapor, and lamp manufacturers have recently made significant progress in developing fluorescent and metal halide lamps which have much more consistent color rendering properties allowing greater flexibility in mixing these two sources without creating disturbing color mismatches. The recent compact fluorescent lamps open up a whole new market for fluorescent sources. These lamps permit design of much smaller luminaries which can compete with incandescent and mercury vapor in the low cost, square or round fixture market which the incandescent and mercury sources have dominated for so long. While generally good, lumen maintenance throughout the lamp lifetime is a problem for some fluorescent lamp types.

Energy Efficient "Plus" Fluorescents[3]

The energy efficient "plus" fluorescents represent the second generation of improved fluorescent lighting. These bulbs are available for replacement of standard 4-foot, 40-watt bulbs and require only 32 watts of electricity to produce essentially the same light levels. To the author's knowledge, they are not available for 8-foot fluorescent bulb retrofit. The energy efficient plus fluorescents require a ballast change. The light output is similar to the energy efficient bulbs, and the two types may be mixed in the same area if desired.

Examples of energy efficient plus tubes include the Super-Saver Plus by Sylvania and General Electric's Watt Mizer Plus.

Energy Efficient Fluorescents System Change

The third generation of energy efficient fluorescents requires both a ballast and a fixture replacement. The standard 2-foot by 4-foot fluorescent fixture, containing four bulbs and two ballasts, requires approximately 180 watts (40 watts per tube and 20 watts per ballast). The new generation fluorescent manufacturers claim the following:

- General Electric — "Optimizer" requires only 116 watts with a slight reduction in light output.
- Sylvania — "Octron" requires only 132 watts with little reduction in light level.
- General Electric — "Maximizer" requires 169 watts but supplies 22 percent more light output.

---

[3] Source: *Fluorescent Lighting—An Expanding Technology*, R. E. Webb, M. G. Lewis, W. C. Turner.

The fixtures and ballasts designed for the third-generation fluorescents are not interchangeable with earlier generations.

Metal halide lamps fall into a lamp efficacy range of approximately 75-125 lumens per watt. This makes them more energy efficient than mercury vapor but somewhat less so than high pressure sodium. Metal halide lamps generally have fairly good color rendering qualities. While this lamp displays some very desirable qualities, it also has some distinct drawbacks including relatively short life for an HID lamp, long restrike time to restart after the lamp has been shut off (about 15-20 minutes at 70°F) and a pronounced tendency to shift colors as the lamp ages. In spite of the drawbacks, this source deserves serious consideration and is used very successfully in many applications.

High pressure sodium lamps introduced a new era of extremely high efficacy (60-140 lumens/watt) in a lamp which operates in fixtures having construction very similar to those used for mercury vapor and metal halide. When first introduced, this lamp suffered from ballast problems. These have now been resolved and luminaries employing high quality lamps and ballasts provide very satisfactory service. The 24,000-hour lamp life, good lumen maintenance and high efficacy of these lamps make them ideal sources for industrial and outdoor applications where discrimination of a range of colors is not critical.

The lamp's primary drawback is the rendering of some colors. The lamp produces a high percentage of light in the yellow range of the spectrum. This tends to accentuate colors in the yellow region. Rendering of reds and greens shows a pronounced color shift. This can be compensated for in the selection of the finishes for the surrounding areas, and, if properly done, the results can be very pleasing. In areas where color selection, matching and discrimination are necessary, high pressure sodium should not be used as the only source of light. It is possible to gain quite satisfactory color rendering by mixing high pressure sodium and metal halide in the proper proportions. Since both sources have relatively high efficacies, there is not a significant loss in energy efficiency by making this compromise.

High pressure sodium has been used quite extensively in outdoor applications for roadway, parking and facade or security lighting. This source will yield a high efficiency system; however, it should be used only with the knowledge that foliage and landscaping colors will be severely distorted where high pressure sodium is the only, or predominant, illuminant. Used as a parking lot source, there may be some difficulty in identification of vehicle colors in the lot. It is necessary for the designer or owner to determine the extent of this problem and what steps might be taken to alleviate it.

Recently lamp manufacturers have introduced high pressure sodium lamps with improved color rendering qualities. However, the improvement in color rendering was not gained without cost—the efficacy of the color-improved lamps is somewhat lower, approximately 90 lumens per watt.

Low pressure sodium lamps provide the highest efficacy of any of the sources for general lighting with values ranging up to 180 lumens per watt. Low pressure sodium produces an almost pure yellow light with very high efficacy, and renders all colors gray except yellow or near yellow. This effect results in no color discrimination under low pressure sodium lighting; it is suitable for use in a very limited number of applications. It is an acceptable source for warehouse lighting where it is only necessary to read labels but not to choose items by color. This source has application for either indoor or outdoor safety or security lighting as long as color rendering is not important.

In addition to these primary sources, there are a number of retrofit lamps which allow use of higher efficacy sources in the sockets of existing fixtures. Therefore, metal halide or high pressure sodium lamps can be retrofitted into mercury vapor fixtures, or self-ballasted mercury lamps can replace incandescent lamps. These lamps all make some compromises in operating characteristics, life and/or efficacy.

Figure 7-3 presents data on the efficacy of each of the major lamp types in relation to the wattage rating of the lamps. Without exception, the efficacy of the lamp increases as the lamp wattage rating increases.

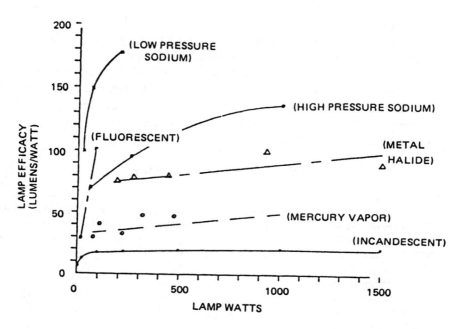

Figure 7-3. Lamp Efficacy
(Does Not Include Ballast Losses)

The lamp efficacies discussed here have been based on the lumen output of a new lamp after 100 hours of operation or the "initial lumens." Not all lamps age in the same way. Some lamp types, such as lightly loaded fluorescent and high pressure sodium, hold up well and maintain their lumen output at a relatively high level until they are into, or past, middle age. Others, as represented by heavily loaded fluorescent, mercury vapor and metal halide, decay rapidly during their early years and then coast along at a relatively lower lumen output throughout most of their useful life. These factors must be considered when evaluating the various sources for overall energy efficiency.

Incandescent Replacement

The most efficacious lamps that can be used in incandescent sockets are the compact fluorescent lamps. The most popular systems are the twin tubes and double twin tubes. These are closer to the size and weight of the incandescent lamp than the earlier type of fluorescent (circline) replacements.

Twin tubes with lamp wattages from 5 to 13 watts provide amounts of light ranging from 240 to 850 lumens. Table 7-1 lists the characteristics of various types of incandescent and compact fluorescent lamps that can be used in the same type sockets.

Table 7-1. Lamp Characteristics

Lamp Type (Total Input Power)*	Lamp Power (W)	Light Output (lumens)	Lamp Life (hour)	Efficacy (lm/W)
100 W (Incandescent)	100	1750	750	18
75 W (Incandescent)	75	1200	750	16
60 W (Incandescent)	60	890	1000	15
40 W (Incandescent)	40	480	1500	12
25 W (Incandescent)	25	238	2500	10
22 W (Fl. Circline)	18	870	9000	40
44 W (Fl. Circline)	36	1750	9000	40
7 W (Twin)	5	240	10000	34
10 W (Twin)	7	370	10000	38
13 W (Twin)	9	560	10000	43
19 W (Twin)	13	850	10000	45
18 W (Solid-State)**	–	1100	7500	61

*Includes ballast losses.
**Operated at high frequency.

The advantages of the compact fluorescent lamps are larger and increased efficacy, longer life and reduced total cost. The cost per 10^6 lumen hours of operating the 75-watt incandescent and the 18-watt fluorescent is $5.47 and $3.29, respectively. This is based upon an energy cost of $0.075 per kWh and lamp costs of $0.70 and $17 and the 75-watt incandescent and 18-watt fluorescent lamps, respectively. The circline lamps were much larger and heavier than the incandescents and would fit in a limited number of fixtures. The twin tubes are only slightly heavier and larger than the equivalent incandescent lamp.

However, there are some fixtures that are too small for them to be employed.

The narrow tube diameter compact fluorescent lamps are now possible because of the recently developed rare earth phosphors. These phosphors have an improved lumen depreciation at high lamp power loadings. The second important characteristic of these narrow band phosphors is their high efficiency in converting the ultraviolet light generated in the plasma into visible light. By proper mixing of these phosphors, the color characteristics (color temperature and color rendering) are similar to the incandescent lamp.

There are two types of compact fluorescent lamps. In one type of lamp system, the ballast and lamp are integrated into a single package; in the second type, the lamp and ballast are separate, and when a lamp burns out it can be replaced. In the integrated system, both the lamp and the ballast are discarded when the lamp burns out.

It is important to recognize when purchasing these compact fluorescent lamps that they provide the equivalent light output of the lamps being replaced. The initial lumen output for the various lamps is shown in Table 7-1.

Lighting Efficiency Options

Several lighting efficiency options are illustrated below: (Refer to Formula 7-2.)

Footcandle Level—The footcandle level required is that level at the task. Footcandle levels can be lowered to one third of the levels for surrounding areas such as aisles. (A minimum 20-footcandle level should be maintained.)

The placement of the lamp is also important. If the luminaire can be lowered or placed at a better location, the lamp wattage may be reduced.

Coefficient of Utilization (Cu)—The color of the walls, ceiling, and floors, the type of luminaire, and the characteristics of the room determine the *Cu*. This value is determined based on manufacturer's literature. The *Cu* can be improved by analyzing components such as lighter colored walls and more efficient luminaires for the space.

Lamp Depreciation Factor and Dirt Depreciation Factor—
These two factors are involved in the maintenance program.
Choosing a luminaire which resists dirt build-up, group relamping and cleaning the luminaire will keep the system in optimum
performance. Taking these factors into account can reduce the
number of lamps initially required.

The light loss factor *(LLF)* takes into account that the lamp
lumen depreciates with time (L_1), that the lumen output depreciates due to dirt build-up (L_2), and that lamps burn out
(L_3). Formula 7-3 illustrates the relationship of these factors.

$$LLF = L_1 \times L_2 \times L_3 \qquad Formula\ (7\text{-}3)$$

To reduce the number of lamps required which in turn
reduces energy consumption, it is necessary to increase the
overall light loss factor. This is accomplished in several ways.
One is to choose the luminaire which minimizes dust build-up.
The second is to improve the maintenance program to replace
lamps prior to burn-out. Thus if it is known that a group relamping program will be used at a given percentage of rated life,
the appropriate lumen depreciation factor can be found from
manufacturer's data. It may be decided to use a shorter relamping period in order to increase (L_1) even further. If a group
relamping program is used, (L_3) is assumed to be unity.

Figure 7-4 illustrates the effect of dirt build-up on (L_2) for
a dustproof luminaire. Every liminaire has a tendency for dirt
build-up. Manufacturer's data should be consulted when estimating (L_2) for the luminaire in question.

Solid-State Ballasts

After more than 10 years of development and 5 years of
manufacturing experience, operating fluorescent lamps at high
frequency (20 to 30 kHz) with solid-state ballasts has achieved
credibility. The fact that all of the major ballast manufacturers
offer solid-state ballasts and the major lamp companies have
designed new lamps to be operated at high frequency is evidence that the solid-state high frequency ballast is now state-of-
the-art.

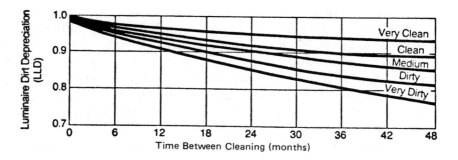

Figure 7-4. Effect of Dirt Build-Up on Dustproof Luminaires for Various Atmospheric Conditions

It has been shown that fluorescent lamps operated at high frequency are 10 to 15 percent more efficacious than 60 Hz operation. In addition, the solid-state ballast is more efficient than conventional ballasts in conditioning the input power for the lamps such that the total system efficacy increase is between 20 and 25 percent. That is, for a standard two-lamp, 40-watt F40 T-12 rapid-start system, overall efficacy is increased from 63 lm/W to over 80 lm/W.

In the past few years, continued development of the product has improved reliability and reduced cost. Today, solid-state ballasts can be purchased for less than $30, and, in sufficiently large quantities, some bids have been less than $20. The industry's growth is evidenced by the availability of ballasts for the 8-foot fluorescent lamp, both slimline and high power, as well as the more common F40 (4-ft) size. In order to be more competitive with initial costs, there are three- and four-lamp ballasts for the F40-type lamps. These multi-lamp ballasts reduce the initial cost per lamp, as well as the installation cost, and are even more efficient than the one- and two-lamp ballast system.

The American National Standards Institute (ANSI) ballast committee has been developing standards for solid-state ballasts for the past few years. The ballast factor is one parameter that will be specified by ANSI. However, solid-state ballasts are available with different ballast factors. The ballast factor is the light output provided by the ballast-lamp system compared to the light output of the lamp specified by the lamp manufacturer. The ANSI ballast factor standard for 40-watt F40 fluorescent lamps is 95 ± 2.5 percent. Because most solid-state ballasts were initially sold on the retrofit market, their ballasts were designed to have a lower ballast factor. Thus, energy was saved not only by the increased efficacy but also by reducing the light output. The thrust was to reduce illumination levels in overlit spaces.

Today, there are solid-state ballasts with a ballast factor exceeding 100 percent. These ballasts are most effectively used in new installations. In these layouts, more light from each luminaire will reduce the number of luminaires, ballasts and lamps, hence reducing both initial and operating costs. It is essential that the lighting designer-engineer and consumer know the ballast factor for the lamp-ballast system. The ballast factor for a ballast also depends upon the lamp. For example, a core-coil ballast will have a ballast factor of 95 ± 2 percent when operating a 40-watt F40 argon-filled lamp and less when operating an "energy saving" 34-watt F40 Krypton-filled lamp. The ballast factor instead will be about 87 ± 2.5 percent with the 34-watt energy saving lamp. Because of this problem, the ANSI standard for the ballast factor for the 34-watt lamps has recently been reduced to 85 percent. Table 7-2 provides some data for several types of solid-state ballasts operating 40-watt and 34-watt F40 lamps and lists some parameters of concern for the consumer.

Table 7-2. Performance of F40 Fluorescent Lamp Systems

| Characteristic | Core-Coil 2 Lamps, T-12 | | — Solid-State Ballasts — | | | | 2 Lamps, T-8 |
	40W	34W	2 Lamps, T-12 40W	34W	4 Lamps, T-12 40W	34W	32W
Power (W)	96	79	72	63	136	111	65
Power Factor (%)	98	92	95	93	94	94	89
Filament Voltage (V)	3.5	3.6	3.1	3.1	2.0	1.6	0
Light Output (lm)	6050	5060	5870	5060	11,110	9250	5820
Ballast Factor	.968	.880	.932	.865	.882	.791	1.003
Flicker (%)	30	21	15	9	1	0	1
System Efficacy (lm/W)	63	64	81	81	82	83	90

The table compares several types of solid-state ballasts with a standard core-coil ballast that meets the ANSI standard with the two-lamp, 40-watt F40 lamp. Notice that the system efficacy of any ballast system is about the same operating a 40-watt or a 34-watt F40 lamp. Although the 34-watt "lite white" lamp is about 6 percent more efficient than the 40-watt "cool white" lamp, the ballast losses are greater with the 34-watt lamp due to an increased lamp current. The lite white phosphor is more efficient than the cool white phosphor but has poorer color rendering characteristics.

Note that the percent flicker is drastically reduced when the lamps are operated at high frequency with solid-state ballasts. A recent scientific field study of office workers in the U.K. has shown that complaints of headaches and eye-strain are 50 percent less under high frequency lighting when compared to lamps operating at 50 cycles, the line frequency of the U.K.

Each of the above ballasts has different factors, which are lower when operating the 34-watt Krypton-filled lamp. Table 7-2 also lists the highest system efficacy of 90 lumens per watt for the solid-state ballast and T-8, 32-watt lamp.

All of the above solid-state ballasts can be used in place of core-coil ballasts specified to operate the same lamps. To determine the illumination levels, or the change in illumination levels, the manufacturer must supply the ballast factor for the lamp type employed. The varied light output from the various systems allows the lighting designer-engineer to precisely tailor the lighting level.

CONTROL EQUIPMENT

Table 7-3 lists various types of equipment that can be components of a lighting control system, with a description of the predominant characteristic of each type of equipment. Static equipment can alter light levels semipermanently. Dynamic equipment can alter light levels automatically over short intervals to correspond to the activities in a space. Different sets of components can be used to form various lighting control systems in order to accomplish different combinations of control strategies.

Table 7-3. Lighting Control Equipment

System	Remarks
STATIC:	
Delamping	Method for reducing light level 50%.
Impedance Monitors	Method for reducing light level 30, 50%.
DYNAMIC:	
Light Controllers	
Switches/Relays	Method for on-off switching of large banks of lamps.
Voltage/Phase Control	Method for controlling light level continuously 100 to 50%.
Solid-State Dimming Ballasts	Ballasts that operate fluorescent lamps efficiently and can dim them continuously (100 to 10%) with low voltage.
SENSORS:	
Clocks	System to regulate the illumination distribution as a function of time.
Personnel	Sensor that detects whether a space is occupied by sensing the motion of an occupant.
Photocell	Sensor that measures the illumination level of a designated area.
COMMUNICATION:	
Computer/Microprocessor	Method for automatically communicating instructions and/or input from sensors to commands to the light controllers.
Power-Line Carrier	Method for carrying information over existing power lines rather than dedicated hard-wired communication lines.

FLUORESCENT LIGHTING CONTROL SYSTEMS

The control of fluorescent lighting systems is receiving increased attention. Two major categories of lighting control are available—personnel sensors and lighting compensators.

Personnel Sensors

There are three classifications of personnel sensors—ultrasonic, infrared and audio.

Ultrasonic sensors generate sound waves outside the human hearing range and monitor the return signals. Ultrasonic sensor systems are generally made up of a main sensor unit with a network of satellite sensors providing coverage throughout the lighted area. Coverage per sensor is dependent upon the sensor type and ranges between 500 and 2,000 square feet. Sensors may be mounted above the ceiling, suspended below the ceiling or mounted on the wall. Energy savings are dependent upon the room size and occupancy. Advertised savings range from 20 to 40 percent.

Several companies manufacture ultrasonic sensors including Novita and Unenco.

Infrared sensor systems consist of a sensor and control unit. Coverage is limited to approximately 130 square feet per sensor. Sensors are mounted on the ceiling and usually directed towards specific work stations. They can be tied into the HVAC control and limit its operation also. Advertised savings range between 30 and 50 percent. (See Figure 7-5.)

Audio sensors monitor sound within a working area. The coverage of the sensor is dependent upon the room shape and the mounting height. Some models advertise coverage of up to 1,600 square feet. The first cost of the audio sensors is approximately one-half that of the ultrasonic sensors. Advertised energy savings are approximately the same as the ultrasonic sensors. Several restrictions apply to the use of the audio sensors. First, normal background noise must be less than 60 dB. Second, the building should be at least 100 feet from the street and may not have a metal roof.

**Figure 7-5. Transformer, Relay and Wide View Infrared Sensor
to Control Lights**
(Photograph courtesy of SensorSwitch)

Lighting Compensators

Lighting compensators are divided into two major groups—switched and sensored.

Switched compensators control the light level using a manually operated wall switch. These particular systems are used frequently in residential settings and are commonly known as "dimmer switches." Based on discussions with manufacturers, the switched controls are available for the 40-watt standard fluorescent bulbs only. The estimated savings are difficult to determine, as usually switched control systems are used to control room mood. The only restriction to their use is that the luminaire must have a dimming ballast.

Sensored compensators are available in three types. They may be very simple or very complex. They may be integrated with the building's energy management system or installed as a stand-alone system. The first type of system is the Excess Light Turn-Off (ELTO) system. This system senses daylight levels and automatically turns off lights as the sensed light level approaches a programmed upper limit. Advertised paybacks for these types of systems range from 1.8 to 3.8 years.

The second type of system is the Daylight Compensator (DAC) system. This system senses daylight levels and automatically dims lights to achieve a programmed room light level. Advertised savings range from 40 to 50 percent. The primary advantage of this system is it maintains a uniform light level across the controlled system area. The third system type is the Daylight Compensator + Excess Light Turn-Off system. As implied by the name, this system is a combination of the first two systems. It automatically dims light outputs to achieve a designated light level and, as necessary, automatically turns off lights to maintain the desired room conditions.

Specular reflectors: Fluorescent fixtures can be made more efficient by the insertion of a suitably shaped specular reflector. The specular reflector material types are aluminum, silver and multiple dielectric film mirrors. The latter two have the highest reflectivity while the aluminum reflectors are less expensive.

Measurements show the fixture efficiency with higher reflectance specular reflectors (silver or dielectric films) is improved by 15 percent compared to a new fixture with standard diffuse reflectors.

Specular reflectors tend to concentrate more light downward with reduced light at high exit angles. This increases the light modulation in the space, which is the reason several light readings at different sites around the fixture are required for determining the average illuminance. The increased downward component of candle power may increase the potential for reflected glare from horizontal surfaces.

When considering reflectors, information should be obtained on the new candle power characteristics. With this information a lighting designer or engineer can estimate the potential changes in modulation and reflected glare.

ENERGY MANAGEMENT

The availability of computers at moderate costs and the concern for reducing energy consumption have resulted in the application of computer-based controllers to more than just industrial process applications. These controllers, commonly called Energy Management Systems (EMS), can be used to control virtually all non-process energy using pieces of equipment in buildings and industrial plants. Equipment controlled can include fans, pumps, boilers, chillers and lights. This section will investigate the various types of Energy Management Systems which are available and illustrate some of the methods used to reduce energy consumption.

The Timeclock

One of the simplest and most effective methods of conserving energy in a building is to operate equipment only when it is needed. If, due to time, occupancy, temperature or other means, it can be determined that a piece of equipment does not need to operate, energy savings can be achieved without affecting occupant comfort by turning off the equipment.

One of the simplest devices to schedule equipment operation is the mechanical timeclock. The timeclock consists of a rotating disk which is divided into segments corresponding to the hour of the day and the day of the week. This disk makes one complete revolution in, depending on the type, a 24-hour or a 7-day period. (See Figure 7-6.)

On and off "lugs" are attached to the disk at appropriate positions corresponding to the schedule for the piece of equipment. As the disk rotates, the lugs cause a switch contact to open and close, thereby controlling equipment operation.

A common application of timeclocks is scheduling office building HVAC equipment to operate during business hours Monday through Friday and to be off all other times. As is

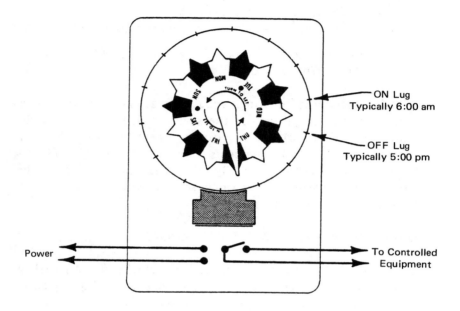

Figure 7-6. Mechanical Timeclock

shown in the following problem, significant savings can be achieved through the correct application of timeclocks.

SIM 7-2

An office building utilizes two 50 hp supply fans and two 15 hp return fans which operate continuously to condition the building. What are the annual savings that result from installing a timeclock to operate these fans from 7:00 a.m. to 5:00 p.m., Monday through Friday? Assume an electrical rate of $0.08/ kWh.

ANSWER

Annual Operation Before Timeclock =
52 weeks X 7 days/week X 24 hours/day = 8736 hours

Annual Operation After Timeclock =
52 × (5 days/week × 10 hours/day) = 2600 hours

Savings = 130 hp × 0.746 kW/hp × (8736 − 2600) hours × $0.08/kWh = $47,600

Although most buildings today utilize some version of a timeclock, the magnitude of the savings value in this example illustrates the importance of correct timeclock operation and the potential for additional costs if this device should malfunction or be adjusted inaccurately. Note that the above example also ignores heating and cooling savings which would result from the installation of a timeclock.

PROBLEMS WITH MECHANICAL TIMECLOCKS

Although the use of mechanical timeclocks in the past has resulted in significant energy savings, they are being replaced by Energy Management Systems because of problems that include the following:

- The on/off lugs sometimes loosen or fall off.

- Holidays, when the building is unoccupied, cannot easily be taken into account.

- Power failures require the timeclock to be reset or it is not synchronized with the building schedule.

- Inaccuracies in the mechanical movement of the timeclock prevent scheduling any closer than ±15 minutes of the desired times.

- There are a limited number of on and off cycles possible each day.

- It is a time-consuming process to change schedules on multiple timeclocks.

Energy Management Systems, or sometimes called electronic timeclocks, are designed to overcome these problems plus provide increased control of building operations.

ENERGY MANAGEMENT SYSTEMS

Recent advances in digital technology, dramatic decreases in the cost of this technology and increased energy awareness have resulted in the increased application of computer-based controllers (i.e., Energy Management Systems) in commercial buildings and industrial plants. These devices can control any-where from one to a virtually unlimited number of items of equipment.

By concentrating the control of many items of equipment at a single point, the EMS allows the building operator to tailor building operation to precisely satisfy occupant needs. This ability to maximize energy conservation, while preserving occupant comfort, is the ultimate goal of an energy engineer.

Microprocessor Based

Energy Management Systems can be placed in one of two broad, and sometimes overlapping, categories referred to as microprocessor-based and mini-computer based.

Microprocessor-based systems can control from 1 to 40 input/outpoints and can be linked together for additional loads. Programming is accomplished by a keyboard or hand-held console, and an LED display is used to monitor/review operation of the unit. A battery maintains the programming in the event of power failure. (See Figures 7-7 and 7-8.)

Capabilities of this type of EMS are generally preprogrammed so that operation is relatively straightforward. Programming simply involves entering the appropriate parameters (e.g., the point number and the on and off times) for the desired function. Microprocessor-based EMS can have any or all of the following capabilities:

- Scheduling
- Duty Cycling
- Demand Limiting
- Optimal Start
- Monitoring
- Direct Digital Control

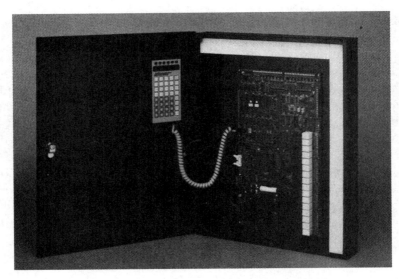

Figure 7-7. Manufacturing a Microprocessor-Based Programmable EMS
(Photograph Courtesy Control Systems International)

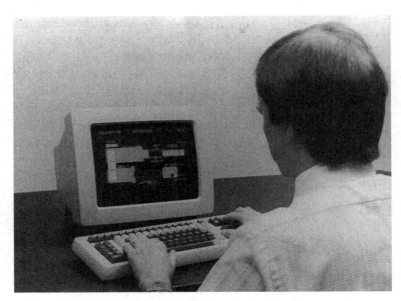

Figure 7-8. Keyboard Used to Access EMS
(Photograph Courtesy Control Systems International)

Scheduling

Scheduling with an EMS is very much the same as it is with a timeclock. Equipment is started and stopped based on the time of day and the day of week. Unlike a timeclock, however, multiple start/stops can be accomplished very easily and accurately (e.g., in a classroom, lights can be turned off during morning and afternoon break periods and during lunch). It should be noted that this single function, if accurately programmed and depending on the type of facility served, can account for the largest energy savings attributable to an EMS.

Additionally, holiday dates can be entered into the EMS a year in advance. When the holiday occurs, regular programming is overridden and equipment can be kept off.

Duty Cycling

Most HVAC fan systems are designed for peak load conditions, and consequently these fans are usually moving much more air than is needed. Therefore, they can sometimes be shut down for short periods each hour, typically 15 minutes, without affecting occupant comfort. Turning equipment off for predetermined periods of time during occupied hours is referred to as duty cycling, and can be accomplished very easily with an EMS. Duty cycling saves fan and pump energy but does not reduce the energy required for space heating or cooling since the thermal demand must still be met.

The more sophisticated EMSs monitor the temperature of the conditioned area and use this information to automatically modify the duty cycle length when temperatures begin to drift. If, for example, the desired temperature in an area is 70° and at this temperature equipment is cycled 50 minutes on and 10 minutes off, a possible temperature-compensated EMS may respond as shown in Figure 7-9. As the space temperature increases above (or below if so programmed) the setpoint, the equipment off time is reduced until, at 80° in this example, the equipment operates continuously.

Duty cycling of fans which provide the only air flow to an

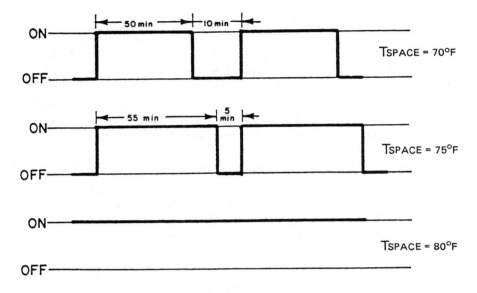

Figure 7-9. Temperature Compensated Duty Cycling

area should be approached carefully to insure that ventilation requirements are maintained and that varying equipment noise does not annoy the occupants. Additionally, duty cycling of equipment imposes extra stress on motors and associated equipment. Care should be taken, particularly with motors over 20 hp, to prevent strating and stopping of equipment in excess of what is recommended by the manufacturer.

Demand Charges

Electrical utilities charge commercial customers based not only on the amount of energy used (kWh) but also on the peak demand (kW) for each month. Peak demand is very important to the utility so that they may properly size the required electrical service and insure that sufficient peak generating capacity is available to that given facility.

In order to determine the peak demand during the billing period, the utility establishes short periods of time called the demand interval (typically 15, 30, or 60 minutes). The billing demand is defined as the highest average demand recorded during any one demand interval within the billing period. (See Figure 7-10.) Many utilities now utilize "ratchet" rate charges. A "ratchet" rate means that the billed demand for the month is based on the highest demand in the previous 12 months, or an average of the current month's peak demand and the previous highest demand in the past year.

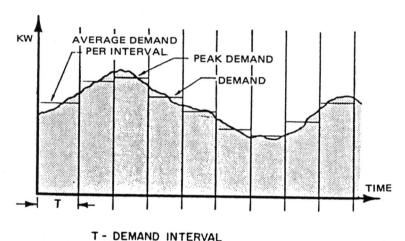

Figure 7-10. Peak Demand

Depending on the facility, the demand charge can be a significant portion, as much as 20%, of the utility bill. The user will get the most electrical energy per dollar if the load is kept constant, thereby minimizing the demand charge. The objective of demand control is to even out the peaks and valleys of consumption by deferring or rescheduling the use of energy during peak demand periods.

A measure of the electrical efficiency of a facility can be found by calculating the load factor. The load factor is defined as the ratio of energy usage (kWh) per month to the peak demand (kW) X the facility operating hours.

SIM 7-3

What is the load factor of a continuously operating facility that consumed 800,000 kWh of energy during a 30-day billing period and established a peak demand of 2000 kW?

ANSWER

$$\text{Load Factor} = \frac{800,000 \text{ kWh}}{2000 \text{ kW} \times 30 \text{ days} \times 24 \text{ hours/day}} = 0.55$$

The ideal load factor is 1.0, at which demand is constant; therefore, the difference between the calculated load factor and 1.0 gives an indication of the potential for reducing peak demand (and demand charges) at a facility.

Demand Limiting

Energy Management Systems with demand limiting capabilities utilize either pulses from the utility meter or current transformers to predict the facility demand during any demand interval. If the facility demand is predicted to exceed the user-entered setpoint, equipment is "shed" to control demand. Figure 7-11 illustrates a typical demand chart before and after the actions of a demand limiter.

Electrical load in a facility consists of two major categories: essential loads which include most lighting, elevators, escalators, and most production machinery; and nonessential ("sheddable") loads such as electric heaters, air conditioners, exhaust fans, pumps, snow melters, compressors and water heaters. Sheddable loads will not, when turned off for short periods of time to control demand, affect productivity or comfort.

To prevent excessive cycling of equipment, most Energy Management Systems have a deadband that demand must drop below before equipmnent operation is restored (see Figure 7-12). Additionally, minimum on and maximum off times and shed priorities can be entered for each load to protect equipment and insure that comfort is maintained.

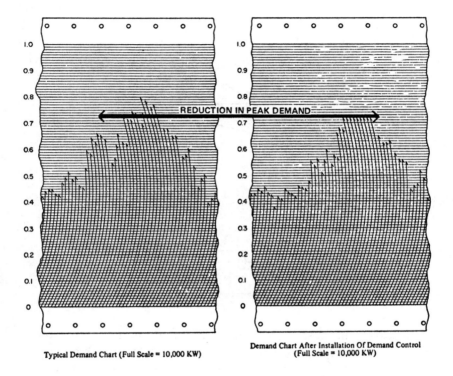

Typical Demand Chart (Full Scale = 10,000 KW)

Demand Chart After Installation Of Demand Control
(Full Scale = 10,000 KW)

Figure 7-11. Demand Limiting Comparison

It should be noted that demand shedding of HVAC equipment in commercial office buildings should be applied with caution. Since times of peak demand often occur during times of peak air conditioning loads, excessive demand limiting can result in occupant discomfort.

Time of Day Billing

Many utilities are beginning to charge their larger commercial users based on the time of day that consumption occurs. Energy and demand during peak usage periods (i.e., summer weekday afternoons and winter weekday evenings) are billed at much higher rates than consumption during other times. This

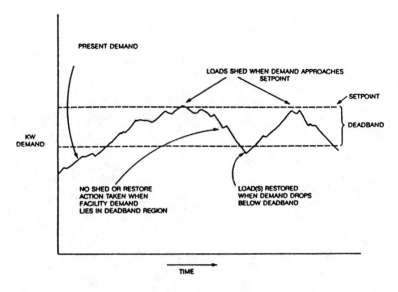

Figure 7-12. Demand Limiting Actions

is necessary because utilities must augment the power production of their large power plants during periods of peak demand with small generators which are expensive to operate. Some of the more sophisticated Energy Management Systems can now account for these peak billing periods with different demand setpoints based on the time of day and day of week.

Optimal Start

External building temperatures have a major influence on the amount of time it takes to bring the building temperature up to occupied levels in the morning. Buildings with mechanical time clocks usually start HVAC equipment operation at an early enough time in the morning (as much as 3 hours before occupancy time) to bring the building up to temperature on the coldest day of the year. During other times of the year when temperatures are not as extreme, building temperatures can be up to occupied levels several hours before it is necessary, and consequently unnecessary energy is used. (See Figure 7-13.)

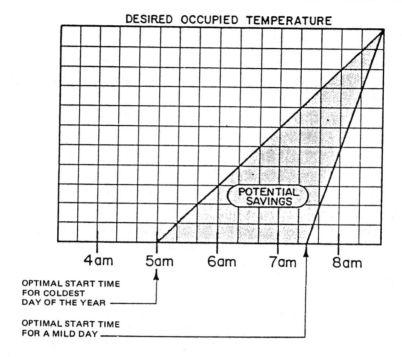

Figure 7-13. Typical Variation in Building Warm-Up Times

Energy Management Systems with optimal start capabilities, however, utilize indoor and outdoor temperature information, along with learned building characteristics, to vary start time of HVAC equipment so that building temperatures reach desired values just as occupancy occurs. Consequently, if a building is scheduled to be occupied at 8:00 a.m., on the coldest day of the year, the HVAC equipment may start at 5:00 a.m. On milder days, however, equipment may not be started until 7:00 a.m. or even later, thereby saving significant amounts of energy.

Most Energy Management Systems have a "self-tuning" capability to allow them to learn the building characteristics. If the building is heated too quickly or too slowly on one day, the start time is adjusted the next day to compensate.

Monitoring

Microprocessor-based EMS can usually accomplish a limited amount of monitoring of building conditions including the following:

- Outside air temperature
- Several indoor temperature sensors
- Facility electrical energy consumption and demand
- Several status input points

The EMS can store the information to provide a history of the facility. Careful study of these trends can reveal information about facility operation that can lead to energy conservation strategies that might not otherwise be apparent.

Direct Digital Control

The most sophisticated of the microprocessor-based EMSs provide a function referred to as direct digital control (DDC). This capability allows the EMS to provide not only sophisticated energy management but also have temperature control of the building's HVAC systems.

Direct digital control has taken over the majority of all process control applications and is now becoming an important part of the HVAC industry. Traditionally, pneumatic controls were used in most commercial facilities for environmental control.

The control function in a traditional facility is performed by a pneumatic controller which receives its input from pneumatic sensors (i.e., temperature, humidity) and sends control signals to pneumatic actuators (valves, dampers, etc.). Pneumatic controllers typically perform a single, fixed function which cannot be altered unless the controller itself is changed or other hardware is added. (See Figure 7-14 for a typical pneumatic control configuration.)

With direct digital control, the microprocessor functions as the primary controller. Electronic sensors are used to measure variables such as temperature, humidity and pressure. This

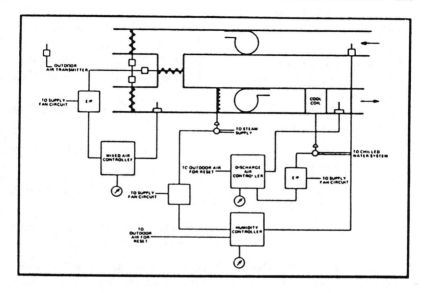

CONVENTIONAL PNEUMATIC CONTROL SYSTEM

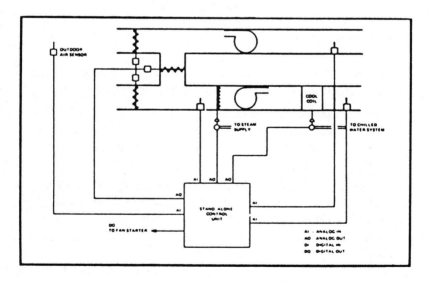

DIRECT DIGITAL CONTROL SYSTEM

Figure 7-14. Comparison of Pneumatic and DDC Controls

information is used, along with the appropriate application program, by the microprocessor to determine the correct control signal, which is then sent directly to the controlled device (valve or damper actuator). (See Figure 7-14 for a typical DDC configuration.)

Direct digital control (DDC) has the following advantage over pneumatic controls:

- Reduces overshoot and offset errors, thereby saving energy.

- Flexibility to easily and inexpensively accomplish changes of control strategies.

- Calibration is maintained more accurately, thereby saving energy and providing better performance.

To program the DDC functions, a user programming language is utilized. This programming language uses simple commands in English to establish parameters and control strategies.

Mini-Computer Based

Mini-computer-based EMS can provide all the functions of the microprocessor-based EMS, as well as the following:
- Extensive graphics
- Special reports and studies
- Fire and security monitoring and detection
- Custom programs

These devices can control and monitor from 50 to an unlimited number of points and form the heart of a building's (or complex's) operations.

Figure 7-15 shows a typical configuration for this type of system.

The "central processing unit" (CPU) is the heart of the EMS. It is a mini-computer with memory for the operating system and applications software. The CPU performs arithmetic and logical decisions necessary to perform central monitoring and control.

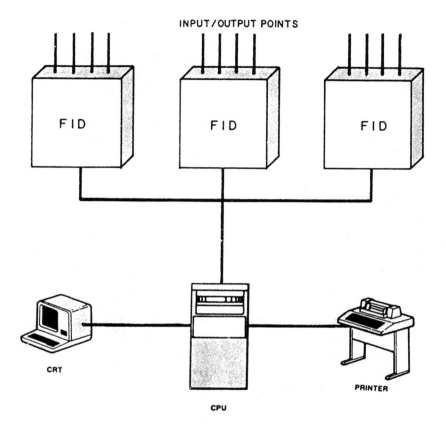

Figure 7-15. Mini-Computer-Based EMS

Data and programs are stored or retrieved from the memory or mass storage devices (generally a disk storage system). The CPU has programmed I/O ports for specific equipment, such as printers and cathode ray tube (CRT) consoles. During normal operation, it coordinates operation of all other EMS components.

A cathode ray tube console (CRT), either color and/or black and white, with a keyboard is used for operator interaction with the EMS. It accepts operator commands, displays data and graphically displays systems controlled or monitored by the

EMS. A "printer" (or printers) provides a permanent copy of system operations and historical data.

A "field interface device" (FID) provides an interface to the points which are monitored and controlled, performs engineering conversions to or from a digital format, performs calculations and logical operations, accepts and processes CPU commands and is capable, in some versions, of stand-alone operations in the event of CPU or communications link failure.

The FID is essentially a microprocessor-based EMS as described in the previous section. It may or may not have a keyboard/display unit on the front panel.

The FIDs are generally located in the vicinity of the points to be monitored and/or controlled and are linked together and to the CPU by a single twisted pair of wires which carries multiplexed data (i.e., data from a number of sources combined on a single channel) from the FID to the CPU and back. In some versions, the FIDs can communicate directly with each other.

Early versions of mini-computer-based EMS used the CPU to perform all of the processing with the FID used merely for input and output. A major disadvantage of this type of "centralized" system is that the loss of the CPU disables the entire control system. The development of "intelligent" FIDs in a configuration known as "distributed processing" helped to solve this problem. This system, which is becoming prevalent today, utilizes microprocessor-based FIDs to function as remote CPUs. Each panel has its own battery pack to insure continued operation should the main CPU fail.

Each intelligent panel sends signals back to the main CPU only upon a change of status rather than continuously transmitting the same value as previous "centralized" systems have done. This streamlining of data flow to the main CPU frees it to perform other functions such as trend reporting. The CPU's primary function becomes one of directing communications between various FID panels, generating reports and graphics and providing operator interface for programming and monitoring.

Features

The primary difference in operating functions of the mini-computer-based EMS is its increased capability to monitor building operations. For this reason, these systems are sometimes referred to as Energy Monitoring and Control Systems (EMCS). Analog inputs such as temperature and humidity can be monitored, as well as digital inputs such as pump or valve status.

The mini-computer-based EMS is also designed to make operator interaction very easy. Its operation can be described as "user friendly" in that the operator, working through the keyboard, enters information in English in a question and response format. In addition, custom programming languages are available so that powerful programs can be created specifically for the building through the use of simplified English commands.

The graphics display CRT can be used to create HVAC schematics, building layouts, bar charts, etc. to better understand building systems operation. These graphics can be "dynamic" so that values and statuses are continuously updated.

Many mini-computer-based EMS can also easily incorporate fire and security monitoring functions. Such a configuration is sometimes referred to as a Building Automation System (BAS). By combining these functions with energy management, savings in initial equipment costs can be achieved. Reduced operating costs can be achieved as well by having a single operator for these systems.

The color graphics display can be particularly effective in pinpointing alarms as they occur within a building and guiding quick and appropriate response to that location. In addition, management of fan systems to control smoke in a building during a fire is facilitated with a system that combines energy management and fire monitoring functions.

Note, however, that the incorporation of fire, security and energy management functions into a single system increases the complexity of that system. This can result in longer start-up time for the initial installation and more complicated troubleshooting if problems occur. Since the function of fire monitoring is critical to building operation, these disadvantages must be weighed against the previously mentioned advantages to determine if a combined BAS is desired.

ELECTRICAL SYSTEM DISTRIBUTION AUDIT

The total power requirement of a load is made up of two components: namely, the resistive part and the reactive part. The resistive portion of a load can not be added directly to the reactive component since it is essentially ninety degrees out of phase with the other. The pure resistive power is known as the watt, while the reactive power is referred to as the reactive volt amperes. To compute the total volt ampere load it is necessary to analyze the power triangle indicated below:

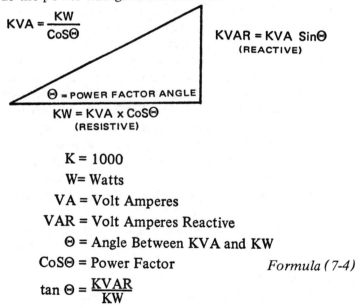

$$KVA = \frac{KW}{CoS\Theta}$$

$$KVAR = KVA\ Sin\Theta \quad \text{(REACTIVE)}$$

$$\Theta = \text{POWER FACTOR ANGLE}$$

$$KW = KVA \times CoS\Theta \quad \text{(RESISTIVE)}$$

$$K = 1000$$
$$W = \text{Watts}$$
$$VA = \text{Volt Amperes}$$
$$VAR = \text{Volt Amperes Reactive}$$
$$\Theta = \text{Angle Between KVA and KW}$$

CoSΘ = Power Factor   *Formula (7-4)*

$$\tan\Theta = \frac{KVAR}{KW}$$

For a balanced 3-phase load

$$\text{Power} = \underbrace{\sqrt{3}\ V_L\ I_L}\ CoS\Theta \quad \text{\textit{Formula (7-5)}}$$

Watts	Volt	Power
	Amperes	Factor

For a balanced 1-phase load

$$P = V_L I_L \cos\Theta \qquad \textit{Formula (7-6)}$$

The standard power rating of a motor is referred to as a horsepower. In order to relate the motor horsepower to a kilowatt (kW) multiply the horsepower by .746 (Conversion Factor) and divide by the motor efficiency.

$$KVA = \frac{HP \times .746}{\eta \times P.F.} \qquad \textit{Formula (7-7)}$$

HP = Motor Horsepower
η = Efficiency of Motor
P.F. = Power Factor of Motor

Motor efficiencies and power factors vary with load. Typical values are shown in Table 7-4. Values are based on totally enclosed fan-cooled motors (TEFC) running at 1800 RPM "T" frame.

Table 7-4

HP RANGE	3-30	40-100
η% at		
½ Load	83.3	89.2
¾ Load	85.8	90.7
Full Load	86.2	90.9
P.F. at		
½ Load	70.1	79.2
¾ Load	79.2	85.4
Full Load	83.5	87.4

Power Factor Efficiency Improvements

The ESEA should collect the following data:
- Plant Power Factor
- Motor nameplate date, type, horsepower, speed, full-load and part-load amperage.
- Nameplate data should be compared to actual running motor amperage.

As indicated in Table 7-4 small, partially-loaded motors contribute to poor power factors and electrical efficiency for buildings and plants.

The ESEA should determine which motors are oversized and may be replaced with a smaller frame size.

A second method to improve the plant or building power factor is to use energy efficient motors. Energy efficient motors are available from manufacturers such as Magnetic. Energy efficient motors are approximately 30 percent more expensive than their standard counterpart. Based on the energy cost it can be determined if the added investment is justified. With the emphasis on energy conservation, new lines of energy efficient motors are being introduced. Figures 7-16 and 7-17 illustrate a typical comparison between energy efficient and standard motors.

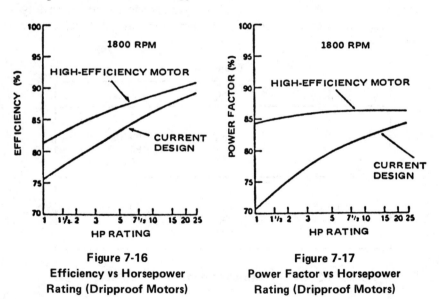

Figure 7-16
Efficiency vs Horsepower
Rating (Dripproof Motors)

Figure 7-17
Power Factor vs Horsepower
Rating (Dripproof Motors)

A third method to improve the power factor is to add capacitor banks to lower the total reactive KVAR. The line current will also be reduced, thus the corresponding $I^2 R$ loss through cables will also be lowered. Table 7-5 can be used to estimate the connective capacitance required.

Table 7-5. Shortcut Method—Power Factor Correction

KW MULTIPLIERS FOR DETERMINING CAPACITOR KILOVARS

ORIGINAL POWER FACTOR IN PERCENTAGE	\ DESIRED POWER-FACTOR IN PERCENTAGE 80	81	82	83	84	85	86	87	88	89	90	91	92	93	94	95	96	97	98	99	100
50	.982	1.008	1.034	1.060	1.086	1.112	1.139	1.165	1.192	1.220	1.248	1.276	1.303	1.337	1.369	1.403	1.441	1.481	1.529	1.590	1.732
51	.936	.962	.988	1.014	1.040	1.066	1.093	1.119	1.146	1.174	1.202	1.230	1.257	1.291	1.323	1.357	1.395	1.435	1.483	1.544	1.688
52	.894	.920	.946	.972	.998	1.024	1.051	1.077	1.104	1.132	1.160	1.188	1.215	1.249	1.281	1.315	1.353	1.393	1.441	1.502	1.644
53	.850	.876	.902	.928	.954	.980	1.007	1.033	1.060	1.088	1.116	1.144	1.171	1.205	1.237	1.271	1.309	1.349	1.397	1.458	1.600
54	.809	.835	.861	.887	.913	.939	.966	.992	1.019	1.047	1.075	1.103	1.130	1.164	1.196	1.230	1.268	1.308	1.356	1.417	1.559
55	.769	.795	.821	.847	.873	.899	.926	.952	.979	1.007	1.035	1.063	1.090	1.124	1.156	1.190	1.228	1.268	1.316	1.377	1.519
56	.730	.756	.782	.808	.834	.860	.887	.913	.940	.968	.996	1.024	1.051	1.085	1.117	1.151	1.189	1.229	1.277	1.338	1.480
57	.692	.718	.744	.770	.796	.822	.849	.875	.902	.930	.958	.986	1.013	1.047	1.079	1.113	1.151	1.191	1.239	1.300	1.442
58	.655	.681	.707	.733	.759	.785	.812	.838	.865	.893	.921	.949	.976	1.010	1.042	1.076	1.114	1.154	1.202	1.263	1.405
59	.618	.644	.670	.696	.722	.748	.775	.801	.828	.856	.884	.912	.939	.973	1.005	1.039	1.077	1.117	1.165	1.226	1.368
60	.584	.610	.636	.662	.688	.714	.741	.767	.794	.822	.849	.878	.905	.939	.971	1.005	1.043	1.083	1.131	1.192	1.334
61	.549	.575	.601	.627	.653	.679	.706	.732	.759	.787	.815	.843	.870	.904	.936	.970	1.008	1.048	1.096	1.157	1.299
62	.515	.541	.567	.593	.619	.645	.672	.698	.725	.753	.781	.809	.836	.870	.902	.936	.974	1.014	1.062	1.123	1.265
63	.483	.509	.535	.561	.587	.613	.640	.666	.693	.721	.749	.777	.804	.838	.870	.904	.942	.982	1.030	1.091	1.233
64	.450	.476	.502	.528	.554	.580	.607	.633	.660	.688	.716	.744	.771	.805	.837	.871	.909	.949	.997	1.058	1.200
65	.419	.445	.471	.497	.523	.549	.576	.602	.629	.657	.685	.713	.740	.774	.806	.840	.878	.918	.966	1.027	1.169
66	.388	.414	.440	.466	.492	.518	.545	.571	.598	.626	.654	.682	.709	.743	.775	.809	.847	.887	.935	.996	1.138
67	.358	.384	.410	.436	.462	.488	.515	.541	.568	.596	.624	.652	.679	.713	.745	.779	.817	.857	.905	.966	1.108
68	.329	.355	.381	.407	.433	.459	.486	.512	.539	.567	.595	.623	.650	.684	.716	.750	.788	.828	.876	.937	1.079
69	.299	.325	.351	.377	.403	.429	.456	.482	.509	.537	.565	.593	.620	.654	.686	.720	.758	.798	.840	.907	1.049
70	.270	.296	.322	.348	.374	.400	.427	.453	.480	.508	.536	.564	.591	.625	.657	.691	.729	.769	.811	.878	1.020
71	.242	.268	.294	.320	.346	.372	.399	.425	.452	.480	.508	.536	.563	.597	.629	.663	.701	.741	.783	.850	.992
72	.213	.239	.265	.291	.317	.343	.370	.396	.423	.451	.479	.507	.534	.568	.600	.634	.672	.712	.754	.821	.963
73	.186	.212	.238	.264	.290	.316	.343	.369	.396	.424	.452	.480	.507	.541	.573	.607	.645	.685	.727	.794	.936
74	.159	.185	.211	.237	.263	.289	.316	.342	.369	.397	.425	.453	.480	.514	.546	.580	.618	.658	.700	.767	.909
75	.132	.158	.184	.210	.236	.262	.289	.315	.342	.370	.398	.426	.453	.487	.519	.553	.591	.631	.673	.740	.882
76	.105	.131	.157	.183	.209	.235	.262	.288	.315	.343	.371	.399	.426	.460	.492	.526	.564	.604	.652	.713	.855
77	.079	.105	.131	.157	.183	.209	.236	.262	.289	.317	.345	.373	.400	.434	.466	.500	.538	.578	.620	.687	.829
78	.053	.079	.105	.131	.157	.183	.210	.236	.263	.291	.319	.347	.374	.408	.440	.474	.512	.552	.594	.661	.803
79	.026	.052	.078	.104	.130	.156	.183	.209	.236	.264	.292	.320	.347	.381	.413	.447	.485	.525	.567	.634	.776
80	.000	.026	.052	.078	.104	.130	.157	.183	.210	.238	.266	.294	.321	.355	.387	.421	.450	.499	.541	.608	.750
81	—	.000	.026	.052	.078	.104	.131	.157	.184	.212	.240	.268	.295	.329	.361	.395	.433	.473	.515	.582	.724
82	—	—	.000	.026	.052	.078	.105	.131	.158	.186	.214	.242	.269	.303	.335	.369	.407	.447	.489	.556	.698
83	—	—	—	.000	.026	.052	.079	.105	.132	.160	.188	.216	.243	.277	.309	.343	.381	.421	.463	.530	.672
84	—	—	—	—	.000	.026	.053	.079	.106	.134	.162	.190	.217	.251	.283	.317	.355	.395	.437	.504	.645
85	—	—	—	—	—	.000	.027	.053	.080	.108	.136	.164	.191	.225	.257	.291	.329	.369	.417	.478	.620

Example: Total kw input of load from wattmeter reading 100 kw at a power factor of 60%. The leading reactive kvar necessary to raise the power factor to 90% is found by multiplying the 100 kw by the factor found in the table, which is .849. Then 100 kw X 0.849 = 84.9 kvar. Use 85 kvar.

Reprinted by permission of Federal Pacific Electric Company.

SUMMARY

The term Energy Management System denotes equipment whose functions can range from simple timeclock control to sophisticated building automation. Two broad and overlapping categories of these systems are microprocessor and mini-computer based.

Capabilities of EMS can include scheduling, duty cycling, demand limiting, optimal start, monitoring, direct digital control, fire detection and security. Direct digital control capability enables the EMS to replace the environmental control system so that it directly manages HVAC operations.

8

The Utility
Energy Audit

Optimizing utility system performance for steam, air, and water
is a very important part of the overall program. This chapter
presents utility and combustion energy audit procedures.

THE COMBUSTION AUDIT

A boiler tuneup should be a high priority on the energy
audit program. The reason being that with a minimal cost, high
operating savings are achieved.

Techniques used to analyze air/fuel ratios, waste heat recovery, and combustion conservation opportunities are presented
in this chapter.

COMBUSTION PRINCIPLES

The boiler plant should be designed and oprated to produce
the maximum amount of usable heat from a given amount of
fuel.

Combustion is a chemical reaction of fuel and oxygen which
produces heat. Oxygen is obtained from the input air which also
contains nitrogen. Nitrogen is useless to the combustion process.
The carbon in the fuel can combine with air to form either CO
or CO_2. Incomplete combustion can be recognized by a low
CO_2 and high CO content in the stack. Excess air causes more
fuel to be burned than required. Stack losses are increased

and more fuel is needed to raise ambient air to stack temperatures. On the other hand, if insufficient air is supplied, incomplete combustion occurs and the flame temperature is lowered.

BOILER EFFICIENCY

Boiler efficiency (E) is defined as:

$$\% E = \frac{\text{Heat out of Boiler}}{\text{Heat supplied to Boiler}} \times 100 \qquad \textit{Formula (8-1)}$$

For steam-generating boiler:

$$\% E = \frac{\text{Evaporation Ratio} \times \text{Heat Content of Steam}}{\text{Calorific Value of Fuel}} \times 100$$
$$\textit{Formula (8-2)}$$

For hot water boilers:

$$\% E = \frac{\text{Rate of Flow from Boiler} \times \text{Heat Output of Water}}{\text{Calorific Value of Fuel} \times \text{Fuel Rate}} \times 100$$
$$\textit{Formula (8-3)}$$

The relationship between steam produced and fuel used is
called the evaporation ratio.

The overall thermal efficiency of the boiler and the various
losses of efficiency of the system are summarized in Figure 8-1.

To calculate dry flue gas loss, Formula (8-4) is used.

$$\text{Flue gas loss} = \frac{K(T-t)}{CO_2} \qquad \textit{Formula (8-4)}$$

where
  K = constant for type of fuel = 0.39 Coke
                                   0.37 Anthracite
                                   0.34 Bituminous Coal
                                   0.33 Coal Tar Fuel
                                   0.31 Fuel Oil
  T = temperature of flue gases in °F
  t = temperature of air supply to furnace in °F
  CO_2 = percentage CO_2 content of flue gas measured volumet-
    rically.

1. Overall thermal efficiency . . . . ─────
2. Losses due to flue gases
   (a) Dry Flue Gas . . . . ─────
   The loss due to heat carried up the stack in dry flue gases can be determined, if the carbon dioxide (CO_2) content of the flue gases and the temperatures of the flue gas and air to the furnace are known.
   (b) Moisture % Hydrogen . . . . ─────
   (c) Incomplete combustion . . . .

3. Balance of account, including radiation and other unmeasured

   ─────

   losses . . . . ─────
                    TOTAL 100%

Figure 8-1. Thermal Efficiency of Boiler

It should be noted that Formula 8-4 does not apply to the combustion of any gaseous fuels, such as natural gas, propane, butane, etc. Basic combustion formulas or nomograms should be used in the gaseous fuel case.

To estimate losses due to moisture, Figure 8-2 is used.
The savings in fuel as related to the change in efficiency is given by Formula 8-5.

$$\text{Savings in Fuel} = \frac{\text{New Efficiency} - \text{Old Efficiency}}{\text{New Efficiency}} \times \text{Fuel Consumption}$$

Formula (8-5)

Figure 8-3 can be used to estimate the effect of flue gas composition, excess air, and stack temperature on Boiler Efficiency.

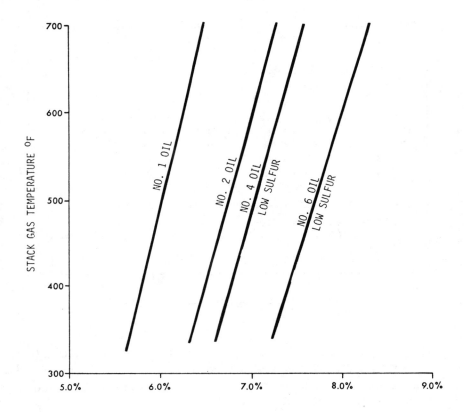

NOTE:
1. The figure gives a simple reference to heat loss in stack gases due to the formation of water in burning the hydrogen in various fuel oils.
2. The graph assumes a boiler room temperature of 80°F.

Figure 8-2. Heat Loss Due to Burning Hydrogen in Fuel
(Source: Instructions For Energy Auditors, Volume 1)

Instructions for use of nomograph (Figure 8-3):

1. Enter the nomograph at the lower horizontal line at the percentage of CO_2 in the flue gas for the fuel being used.

2. Enter the lower left-hand vertical part of the nomograph at the percentage O_2 in the flue gas and proceed horizontally right to the intersection of the plotted curved line.

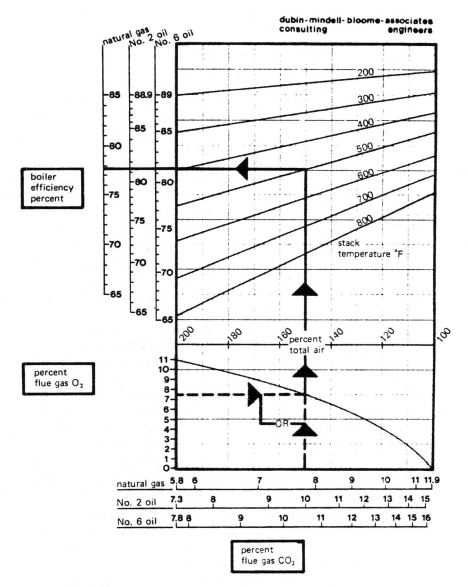

**Figure 8-3. Effect of Flue Gas Composition and Temperature
on Boiler Efficiency**
*(Source: Guidelines For Saving Energy in Existing Buildings—
Engineers, Architects and Operators Manual, ECM-2)*

3. Proceed vertically upward at this intersection to the stack temperature line.

4. Proceed horizontally left at this intersection and read the boiler efficiency corresponding to the fuel used.

Primary and secondary air should be allowed to enter the combustion chamber only in regulated quantities and at the correct place. Defective gaskets, cracked brickwork, broken casings, etc. will allow uncontrolled and varying quantities of air to enter the boiler and will prevent accurate fuel/air ratio adjustment. If spurious stack temperature and/or oxygen content readings are obtained, inspect the boiler for air leaks and repair all defects before a final adjustment of the fuel/air ratio.

When substantial reductions in heating load have been achieved, the firing rate of the boiler may be excessive and should be reduced. Consult the firing equipment manufacturer for specific recommendations. (A reduced firing rate in gas and oil burners may require additional bricking to reduce the size and shape of the combustion chamber.)

Use Figure 8-3 to determine the optimum fuel/air ratio for any given combination of circumstances. Indicators of maximum combustion efficiency are stack temperature, percentage CO_2 and percentage O_2.

Devices are available which continuously measure CO_2 and stack temperature to produce a direct reading of boiler efficiency. These indicators provide boiler operators with the requisite information for manual adjustment of boiler fuel/air ratio. They are suitable for smaller installations or buildings where money for investment in capital improvements is limited. A more accurate measure of combustion efficiency, however, is obtained by an analysis of oxygen content rather than of other gases such as carbon dioxide and carbon monoxide. As shown in Figure 8-3, the cross checking of O_2 concentrations is useful in judging burner performance more precisely. Due to the increasing utilization of multifuel boilers, however, O_2 analysis is the single most useful criterion for all fuels since the O_2 total air ratio varies only within narrow limits.

For larger boiler plants, consider the installation of an automatic continuous oxygen analyzer with "trim" output that will adjust the fuel/air ratio to meet changing stack draft and load conditions. Most boilers can be modified to accept an automatic fuel/air mixture control by a flue gas analyzer, but a gas analyzer manufacturer should be consulted for each particular installation to be sure that all other boiler controls are compatible with the analyzers.

It is important to note that some environmental protection laws might place a higher priority on visible stack emissions than on efficiency and optimization of fuel combustion, especially where fuel oil is burned. The effect of percent total air on smoke density might prove to be an overriding consideration and limit the approach to minimum excess air. All applicable codes and environmental statutes should be checked for compliance.

PREHEAT COMBUSTION AIR AND HEAVY FUEL OIL TO INCREASE BOILER EFFICIENCY

Preheating the primary and secondary air will reduce its cooling effect when it enters the boiler combustion chamber, thus increasing the efficiency of the boiler as indicated in Figure 8-4. It will also promote a more intimate mixing of fuel and air which will further improve efficiency. Waste heat from flue gases, blowdown, condensate, hot wells, etc. may be used to preheat combustion air and/or oil, either in the storage tanks (low sulfur oil requires continuous heating to prevent wax deposits) or at the burner nozzle.

A waste heat exchange directly from flue gases to combustion air using static tubular, plate, or rotary exchangers can be implemented. Heat exchange may also be made indirectly through run-around coils in the stack and combustion air duct. In most boiler rooms, air is heated incidentally by hot boiler and pipe surfaces and rises to collect below the ceiling. Use this air directly as preheated combustion air by ducting it down to the firing level and directing it into the primary and secondary air inlets.

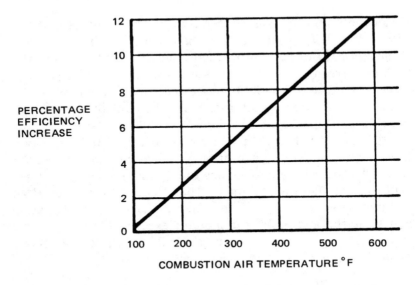

Figure 8-4. Efficiency Increase with Pre-Heated Air

Preheating combustion air has the following advantages:
- Flame temperature is raised, thus permitting air increase in boiler output.
- Higher flame temperature reduces excess air requirements.
- Dual firing is made simpler.

As indicated by Figure 8-4, boiler efficiency will increase by approximately 2 percent for each 100°F added to combustion air temperature. Oil must be preheated to at least the following temperatures to obtain complete atomization:

No. 4 oil — 135°F
No. 5 oil — 185°F
No. 6 oil — 210°F

Heating beyond these temperatures will increase efficiency, but care must be taken not to overheat, as vapor-locking could cause flame-outs. The increased efficiency obtained by preheating oil could be as high as 3 percent but depends on the particular constituents of the oil.

In doing the analysis, obtain the manufacturer's recommendations on preheated fuel and air for the particular equipment being considered. Obtain the fuel oil dealer's recommendations on the preheating levels most appropriate for the fuel to be used. (Combustion air can be preheated up to 600°F for pulverized fuels and up to 350°F for stoker-fired coal, oil, and gas. The upper temperature limit is determined by the design and materials of the firing equipment.)

REPLACE EXISTING BOILERS WITH MODULAR BOILERS

Heating boilers are usually designed to operate at maximum efficiency only when producing their rated output of Btu. As shown in Figure 8-5, however, most boilers operate at 60 percent or less of capacity for 90 percent of the heating season, resulting in significant boiler inefficiencies and wasted fuel.

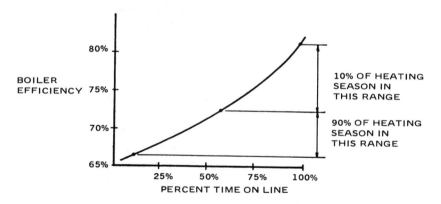

Figure 8-5. Effects of Boiler Cycling

This waste can be diminished but not eliminated by high-low firing systems in large capacity boilers, as shown in Figure 8-6.

A generally superior means of meeting a fluctuating boiler load demand is a system of modular boilers which can be fired independently. Each small-capacity unit has a relatively low thermal inertia (giving rapid response and low heat-up and cool-down losses) and will either be firing at a maximum efficiency can be improved from 68 to 75 percent in a typical installation where single-unit, large-capacity boilers are replaced by any modular boilers. This represents a saving of approximately 9 percent of the present yearly fuel consumption of most commercial buildings.

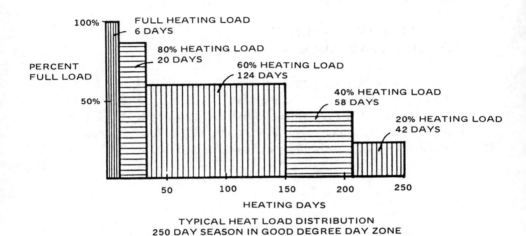

Figure 8-6. Seasonal Heat Load Distribution

To increase seasonal efficiency, it is advisable to replace a single boiler with an array of smaller modular boilers. Each module would be fired on demand at 100 percent capacity, with load fluctuations being met by firing more or fewer boilers. Especially where the present boiler plant has deteriorated to the point where it is at or near the end of its useful life, it is often worthwhile to consider replacement with modular boilers sized to meet the reduced heating load resulting from other ECOs.

Note: Modular boilers are particularly effective in buildings with intermittent short-time occupancy, such as churches. They provide rapid warm-up for occupied periods and low standby losses during extended unoccupied periods.

To calculate the potential energy savings due to this change, refer to Figure 8-6 to determine the seasonal efficiency of the proposed modular boiler installation.

MAINTAIN FUEL BURNING EQUIPMENT
AND HEAT TRANSFER SURFACES

Fuel burning equipment allowed to become dirty and out of adjustment becomes increasingly inefficient with continued usage. Likewise, both fire-side and water-side heat transfer surfaces become less and less effective if allowed to become fouled by products of combustion, scaling, and other impurities. All heat not properly transferred is discharged through the stack.

After reducing the building and distribution heating load, clean and/or replace dirty oil nozzles, oversized or undersized nozzles, fouled gas parts, and improperly sized combustion chambers. Reduce nozzle sizes and modify combustion chambers for proper combustion.

The condition of the heat transfer surface directly affects heat transfer from the combustion chamber and/or hot gases. Keep the fire-side of the heat transfer surface clean and free from soot or other deposits and the air- and water-sides clean and free of scale deposits. Remove deposits by scraping where they are accessible, by chemical treatment, or by a combination. In the case of steam boilers, once the water-side of the boiler is clean, institute correct water treatment and blowdown to maintain optimum heat transfer conditions.

REDUCE BLOWDOWN LOSSES

The purpose of blowing down a boiler is to maintain a low concentration of dissolved and suspended solids in the boiler water and to remove sludge in the boiler to avoid priming and carryover. There are two principal types of blowdown intermittent—manual blowdown and continuous blowdown. Manual blowdown (or sludge blowdown) is necessary for the operation of the boiler regardless of whether continuous blowdown is being used. The frequency of manual blowdown will depend on the volume of solids in the boiler makeup water and the type of water treatment used. While continuous blowdown requires a steady supply of additional energy (because the makeup water must be heated), these losses can be minimized with automatic blowdown control and heat recovery systems.

Install automatic blowdown controls to monitor the conductivity and pH of the boiler water allowing the boiler to blow down only when required to maintain acceptable water quality. Further savings can be realized by piping the blowdown water through a heat exchanger or through a flash tank with a heat exchanger.

To calculate the potential savings:

1. Determine the blow-down rate and calculate the total heat available from blowdown.

2. Compute the heat to be recovered by using a heat exchanger and/or by adding a flash tank.

BURNERS

The choice of a burner is critical to the whole boiler efficiency operation. The basic requirement of an oil burner is that it change the oil into tiny particles thus exposing the greatest surface area of combustible materials in the shortest possible time. Some burners atomize the oil better than others.

Another important aspect is that the burner have the same operating range or turndown ratio as the boiler. Losses of up to 20 percent in fuel consumption may be occurring when a poor turndown ratio burner is matched against a fluctuating steam load. Burners and associated control systems should be able to modulate through the whole range of output called for by the facility.

Air-atomizing burners are considerably more efficient than steam-atomizing burners, due primarily to the relatively higher O_2 content of the fuel at the instant of combustion.

FLAKY PRODUCTS AND SERVICES

As with any new technology, care should be given to "fly-by-night con artists." The market place will clean itself, but in the meantime many people will be hurt. The example of the "unturned automobile" should not be overlooked. If the automobile was "unturned" and fitted with an energy saving carburetor, the end result may be a savings in gasoline. The question asked is whether the savings is the result of the carburetor or a tune-up which had to be done after the unit was installed.

COMPRESSED AIR AUDIT

Air leaks are a major energy loss as indicated in Table 9-3, Chapter 9. Doubling the size of air leak increases the loss four times, as illustrated in Figure 8-7.

The energy audit should determine pressure requirements of each user. If the pressure of the distribution system can be lowered savings will be realized, as illustrated by Figure 8-8. If only one or two users require a higher pressure, it may be desirable to purchase a smaller compressor for these users.

A third area to check is the temperature of the incoming air. The lower the inlet air temperature, the greater the volume of air that can be delivered at room temperature. Thus the installation of a manual inlet damper may be justified. This would permit use of outside air during winter and inside air during summer.

INSULATION

Savings as a result of using the optional economic insulation thickness has been estimated as 1,400 trillion Btus.

Figure 8-9 illustrates minimum recommended pipe size insulation for each pipe diameter.

Several manufacturers offer access to computer program simulation by use of a touch dial telephone and an assigned user number. These programs can calculate economic thickness

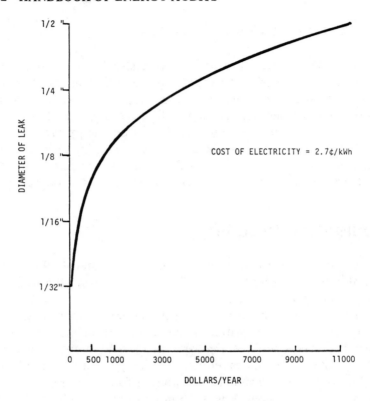

Figure 8-7. Cost of Air Leaks at 100 PSI
(Source: Instructions For Energy Auditors, Volume II)

for tanks as well as equipment and piping. To use these programs the user dials the computer telephone number and then talks to a computer by touching numbers on the telephone. In a simulated voice the computer transmits the economic thickness. The detailed analysis is given to the user by a local sales representative.

The primary function of insulation is to reduce the loss of energy from a surface operating at a temperature other than ambient. The economic use of insulation reduces plant operating expenditures for fuel, power, etc.; improves process efficiency; increases system output capacity; or may reduce the required capital cost.

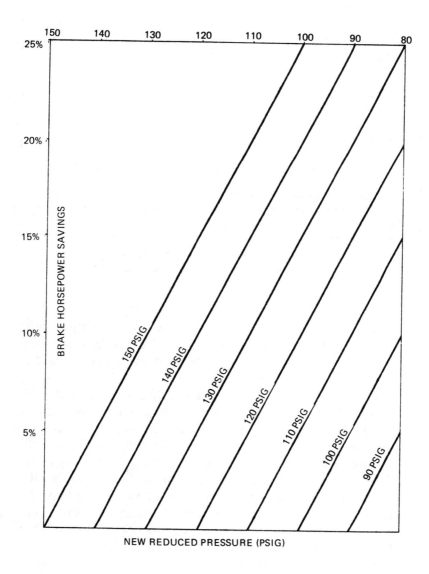

Figure 8-8. Savings with Reduction in Pressure
(Source: Instructions for Energy Auditors, Volume II)

Piping System	Temperature Range °F	Insulation Thickness Inches
Heating		
High pressure steam	306 to 400	1.5 – 2.0
Medium pressure steam	251 to 305	1.0 – 1.5
Low pressure steam	up to 250	1.0
Condensate	190 to 220	1.0
Hot water	up to 200	1.0
Hot water	over 200	1.0
Cooling		
Chilled water	40 to 60	.75 – 1.0
Refrigerant and Brine	below 32	1.0 – 1.5

A WORD OF WARNING: Make absolutely sure that the pipe or vessel to be insulated is properly primed with zinc or silicone coating before installing the insulation.

Figure 8-9. Minimum Piping Insulation
(Source: Instructions for Energy Auditors, Volume II)

There are two costs associated with the insulation type chosen: a cost for the insulation itself, and a cost for the energy lost through this thickness. The total cost for a given period is the sum of both costs.

The optimum economic thickness is that which provides the most cost-effective solution for insulating and is determined when total costs are a minimum. Since the solution calls for the sum of the lost energy and insulation investment costs, both costs must be compared in similar terms. Either the cost of insulation must be estimated for each year and compared to the average annual cost of lost energy over the expected life of the insulation, or the cost of the expected energy loss each year must be expressed in present dollars and compared with the total cost of the insulation investment. The former method, making an annual estimate of the insulation cost and comparing it to the average expected annual cost of lost energy, is the method used in this analysis.

9

Maintenance And Energy Audits

An audit for a good preventive maintenance program as well as good housekeeping methods is essential. Probably no single one area offers the best rate of return and is the most overlooked and underemphasized area. This chapter will illustrate both the administrative and technical areas that make up a good preventive maintenance program.

WORK ASSIGNMENTS

Each major item of equipment must show a history of maintenance and repair. A procedural system of indexing scheduled work and quality control should be established by the supervisor in conjunction with the company's standards of performance.

Personnel assigned to the maintenance control system must be made familiar with all work items. Thus assignments are to be regularly rotated so as to familiarize each man with the equipment.

TRAINING

Periodically, personnel will be requested to attend manufacturers' seminars on maintenance methods for physical plant, building, kitchen equipment, etc.

Typical Manufacturer	*Equipment Type*
A — Sellers Manufacturing Company —	Boilers
B — Hobart Manufacturing Company —	Kitchen Equipment, Dishwashers, etc.
C — Gaylord Manufacturing Company —	Kitchen Exhaust Hoods
D — Vogt Manufacturing Company —	Ice Machines
E — Traulsen —	Reach-In, Pass-Through Refrigeration
F — Groen —	Steam Kettles, Tilting Fry Pans
G — Etc. —	

To assure skilled maintenance personnel and maintenance supervisors, an apprentice mechanics training program should be initiated. The maintenance supervisor will be responsible for the work progression and technical training of the apprentice.

PREVENTIVE MAINTENANCE PROCEDURES

The preventive maintenance (PM) program is a method of budgeting and controlling maintenance expense. It pinpoints problem areas, it helps avoid repetitive maintenance, excessive parts replacement, and purchasing errors. Thus, money spent on a well-planned system of preventive maintenance reduces profit loss due to breakdown, emergency work, and related parts failures.

In order to introduce controls, the PM program must be effective but very simple to avoid assigning administrative chores to maintenance for recordkeeping, etc. When maintenance fills out a simplified work ticket, illustrated in Figure 9-1, the data acquired helps pinpoint costs to accomplish the following:

- Show areas of high cost.
- Change criteria of new construction to reduce high-cost areas.
- Set incentive goals for satisfaction of work.
- Eliminate high-priced skilled labor performing mediocre, unskilled chores.
- Point out high-cost areas to obtain help from qualified technicians, controllers, etc.

NAME: _____

BILLET #: _____

SHIFT: _____

	OPERATION #	HOURS	CODE
1			
2			
3			
4			
5			
6			
7			
8			
OVER-TIME			

REMARKS: _____

_____

_____

_____

CODES: A — APPRENTICE
MJ — MECHANIC (JUNIOR)
M — MECHANIC
S — SUPERVISOR
CL — CLERK

Code Description

AM	Administration
AC	Air Conditioning
B	Boilers
C	Carts
D	Malicious Damage
EP	Electrical Power
EL	Lighting
EV	Elevators
G	Building Maintenance General
H	Heating
K	Kitchen Equipment
LS	Landscaping & Site Work
M	Miscellaneous
P	Plumbing
PM	Preventative Maintenance
R	Refrigeration
T	Supervisor's Technical Time
V	Ventilation
1	Boiler Room
2	Dry Storage
3	Cafeteria
4	Restrooms & Locker Rooms
5	Assembly Area
6	Dishroom
7	Gift Shop
8	Loading Dock
9	General Storage Area
10	General Offices
11	Miscellaneous

Figure 9-1. Simplified Work Ticket

The coding on the ticket will enable study of the shift for more effective coverage of the operation after evaluation of the time cards. Information on time tickets, such as the nature of breakdown and what action was taken to correct it, is only valuable to supervisory personnel for evaluation of satisfactory performance of maintenance duties.

The information needed to reduce costs in PM is compiled from data collected from high-cost areas. This appears in the labor required to effectively maintain the operation and the labor needed for an effective PM program.

A typical coding system is as follows:

AM —   **Administration** —is directed to the maintenance supervisor so as to pinpoint his administrative duties vs. his technical supervision.

AC —   **Air-Conditioning** —is to point out costs in this area to take corrective steps in future program criteria.

B —   **Boiler** —would pertain to breakdown, lack of PM, etc. It can be directly attributed to that area of time required to maintain service.

C —   **Carts (Rolling Stock)** —would be maintenance of casters, modules, baker's racks, dunnage racks, portable mop sinks, etc. It would enable us to pinpoint areas such as specifying heavier duty casters, welding in key point areas, etc.

D —   **Malicious Damage** —would include mistreating equipment (carts, kitchen equipment, etc.)

EP —   **Electrical Power** —would entail the following: from the service entrance, main disconnect, electricity to source of lighting, power to all equipment, etc.

EL —   **Lighting** —will encompass the area of lamp replacement and maintenance of the lighting system throughout the operation.

EV —   **Elevators** —where applicable, will deal strictly in the area that is directly pertaining to the satisfactory function of the elevator.

G —   **Building Maintenance (General)** —will cover the areas of painting, tile replacement, roof repair, windows, etc.

H —   **Heating** —will deal in the areas of what means the building is being heated, such as steam, HVAC units, space heaters, etc.

K —   **Kitchen Equipment** —would entail all equipment which includes ranges, conveyors, dishwashers, etc.

LS —   **Landscaping & Site Work** —would encompass the exterior of the building such as lawn, trees, sprinkler system, paving and striping of roads and lots.

M —   **Miscellaneous** —will be used in areas that have not been covered by defined codes.

P —   **Plumbing** —will deal with the areas of water, sewer, industrial wastes, grease traps, septic tanks, etc.

PM —   **Preventive Maintenance** —will cover time and location spent on preventive maintenance program so as to pinpoint high PM areas.

R —   **Refrigeration** —will cover the areas of maintaining compressors, condensers, evaporative coils on all walk-ins, reach-throughs, and pass-throughs.

T —   **Supervisor's Technical Time** —will cover the amount of time actually spent in supervising maintenance in the field.

V —   **Ventilation** —will cover the areas of the supply air system and exhaust.

To further pinpoint costs numbers should be assigned to descriptive areas. For example:

1. Boiler Room
2. Dry Storage
3. Cafeteria
4. Restroom and Locker room
5. Assembly Areas
6. Dishroom
7. Gift Shop
8. Loading Dock
9. General Storage Areas
10. General Offices
11. Miscellaneous

The purpose for setting up these codes is to shorten the time for filling out the time tickets, which could be a time-consuming and meaningless task when the ultimate goal is to reduce costs and pinpoint high-cost areas. Naturally, as the system is introduced, other codes will be initiated to cover areas that have not been covered in the start-up of the preventive maintenance program.

To keep personality out of the PM program and to reduce administrative chores of the mechanics, helpers, etc., each maintenance supervisor or mechanic assigned (whichever the case may be) will assign a billet number to each maintenance mechanic, helper, etc. The explanation of these codes is as follows:

A	Apprentice	A	A-1 _____ Series
MJ	Mechanic (Junior)	MJ	MJ-2 _____ Series
M	Mechanic	M	M-3 _____ Series
S	Supervisor	S	S-4 _____ Series
CL	Clerk	CL	CL-5 _____ Series

Example: John Smith: M-3 _____ (will be mechanic's billet #). In case John Smith leaves the company, the new employee will be assigned the same billet number for payroll purposes, etc.

The time ticket attached covers an eight-hour shift, but an area has been designated for overtime which will require an explanation from _____ .

The steps required for a fully encompassing preventive maintenance program are described in this section. Each step must be performed initially and then added to and revised as new equipment is purchased or existing equipment requires more frequent maintenance.

The initial organization and subsequent administration of the program is the responsibility of the supervisor of maintenance.

PREVENTIVE MAINTENANCE SURVEY

This survey is made to establish a list of all equipment on the property that requires periodic maintenance and the maintenance that is required. The survey should list all items of equipment according to physical location. The survey sheet should list the following columns:

1. Item
2. Location of Item
3. Frequency of Maintenance
4. Estimated Time Required for Maintenance
5. Time of Day Maintenance Should Be Done
6. Brief Description of Maintenance To Be Done

PREVENTIVE MAINTENANCE SCHEDULE

The preventive maintenance schedule is prepared from the information gathered during the survey. Items are to be arranged on schedule sheets according to physical location. The schedule sheet should list the following columns:

1. Item
2. Location of Item
3. Time of Day Maintenance Should Be Done
4. Weekly Schedule with Double Columns for Each Day of the Week (one column for "scheduled" and one for "completed")
5. Brief Description of Maintenance To Be Done
6. Maintenance Mechanic Assigned To Do the Work

USE OF PREVENTIVE MAINTENANCE SCHEDULE

At some time before the beginning of the week, the supervisor of maintenance will take a copy of the schedule. The copy that the supervisor prepares should be available in a three-ring notebook. He will go over the assignments in person with each mechanic.

After the mechanic has completed the work, he will note this on the schedule by placing a check under the "completed" column for that day and the index card system for cross-checking the PM program.

The supervisor of maintenance or the mechanic will check the schedule daily to determine that all work is being completed according to the plan. At the end of the week, the schedule will be removed from the book and checked to be sure that all work was completed. It will then be filed.

USE OF THE SCHEDULE TO RECORD REPAIRS

Any repairs or replacement of parts on a particular piece of equipment should be noted on the preventive maintenance schedule. The work done should be written in the weekly schedule section or reference should be made to an attached sheet if more space is necessary. This will provide a history of repairs or replacements on each piece of equipment.

The supervisor of maintenance or mechanic should analyze these schedules twice a year to determine if certain pieces of equipment are requiring more than acceptable maintenance and if replacement of the piece of equipment is necessary.

Figure 9-2 illustrates a form used by the supervisor of maintenance, lead mechanic, etc., for accumulating cost and labor information on a weekly, monthly, and yearly basis. The accumulated information pinpoints high-cost areas, preventive maintenance labor, etc., plus the necessary information for yearly budgeting and other purposes.

The index card system illustrated in Figure 9-3 will become a source of data collecting plus a cross check on preventive maintenance schedules and the analyzing time needed to perform PM work for future maintenance schedules. When the initial program goes into effect, it will require estimating the time required to perform PM on equipment. The data collected will also compile a record of which type motor belts, filters, etc., will be needed to reduce inventory, etc.

The card will be placed in a waterproof enclosure and attached to or located near the equipment which will require preventive maintenance. This will eliminate PM being performed in the office, since the mechanic will fill out the required information listed on the index card and will be responsible if the PM work was not performed. It would further help the supervisor of maintenance or lead mechanic to evaluate the mechanic's performance. See Figure 9-4.

PREVENTIVE MAINTENANCE TRAINING

The supervisor of maintenance or mechanic is responsible for assisting department heads in the training of employees in handling, daily care, and the use of equipment. When equipment is mishandled, he must take an active part in correcting this through training.

SPARE PARTS

All too often equipment is replaced with the exact model as presently installed. Excellent energy conservation opportunities exist in upgrading a plant by installing more efficient replace-

REMOVED/EXIST. INVENTORY	
AIR CONDITIONING	
BOILERS	
CARTS	
ELECTRICAL	
LIGHTING	
BLDG. MAINTENANCE (GEN.)	
HEATING	
KITCHEN EQUIPMENT	
LAND & SITE WORK	
MISCELLANEOUS	
PLUMBING	
PREVENTATIVE MAINTENANCE	
REFRIGERATION	
VENTILATION	
LABOR	
OVERTIME	

ACTUAL MAINTENANCE COST EXPENDITURES

Weekly

TOTAL COST OF ESTIMATED MAINTENANCE EXPENDITURE.........

Figure 9-2. Maintenance Cost Expenditure Form

DAILY   WEEKLY   MONTHLY   6 MONTHS   YEARLY

MECHANIC ASSIGNED FOR COMPLETION
BILLET #   AMOUNT OF TIME   DATE

Figure 9-3. Maintenance Index Card System

	DAILY	WEEKLY	MONTHLY	6 MONTHS	YEARLY
	BILLET #		AMOUNT OF TIME	REMARKS	
JAN					
FEB					
MAR					
APR					
MAY					
JUN					
JUL					
AUG					
SEP					
OCT					
NOV					
DEC					

Figure 9-4. Recording Performance Comments Form

ment parts. Consideration should be given to the following:
- Efficient line motor to replace standard motors
- Efficient model burners to replace obsolete burners
- Upgrading lighting systems.

How many times are steam traps replaced with a size corresponding to the pipe thread size? Instead of this energy inefficient procedure, before a steam trap is replaced the correct orifice size should be determined. In this way the steam trap will be checked periodically for correct sizing. When a discharge pipe needs to be replaced because it has corroded, a check should be made to determine if a larger size diameter pipe should be used as its replacement. The larger diameter pipe reduces pipe friction losses, thus saving energy.

EQUIPMENT MAINTENANCE

When equipment is properly maintained energy will be saved. This section contains representative equipment and types of maintenance checks to be performed. In addition to equipment checks, leaks in steam, water, air and other utilities should be made and uninsulated or damaged insulation, furnace refractory damages, etc. should be recorded and corrected.

Figures 9-5 and 9-6 illustrate maintenance survey and log book forms respectively.

EQUIPMENT PM AND OPERATIONS

BOILERS

Operating and maintenance procedures depend on the type of boiler, the fuel used, and the manufacturer's instructions. Permanent records should be kept covering all inspections, testing, and servicing.

A general maintenance checklist is illustrated in Figure 9-7. A specific form similar to this figure should be incorporated into the overall PM program based on the details of the unit in operation.

Date	Operations Request	Maintenance				
		Date	Code	Billet #	Action Taken	

Figure 9-5. Operations Maintenance Log Book

MAINTENANCE DEPARTMENT

OPERATION NO.

ITEM	LOCATION OF ITEM	CODE	FREQUENCY OF MAINTENANCE	EST. TIME REQUIRED	TIME OF DAY	DESCRIPTION OF MAINTENANCE

Figure 9-6. Preventive Maintenance Survey Form

System	Daily Requirements	Weekly	Monthly	Annual
Blowdown and Water Treatment	• Check that blowdown valve does not leak. • Make sure blowdown is not excessive.		• Make sure that solids are not built-up.	
Exhaust Gases	• Check temperature at two different firings	• Measure exhaust gas temperature and composition at selected firings and adjust to recommended values.	• Same as weekly. • Compare with readings of previous months	• Same as weekly. • Record reference data
Burner	• Check controls are operating properly. • Burner may need cleaning several times daily if #6 fuel is used.	• Clean burner pilot pilot assemblies. • Check condition of spark gap, electrode, burner.	• Same as weekly.	• Same as weekly. • Clean and recondition
Feedwater Systems	• Check & correct unstable water level. • Causes of unstable conditions: contaminants, overload, malfunction.	• Check control by stopping feedwater pump and allow control to stop fuel flow.		• Clean condensate receivers, de-aeration system. • Check pumps.
Steam Pressure	• Check for excessive loading on boiler which will cause excessive variations in pressure.			
Air Temperature in Boiler Rooms	• Check that temperature in boiler room is within acceptable range			

Relief Valve	• Check if relief valve leaks.		• Remove and recondition.
Boiler Operating Characteristics	• Observe flame failure system & characteristics of flame.		
Combustion Air Supply		• Check that adequate openings exist for combustion air inlet. • Clean inlet if fouled.	
Fuel System		• Check pumps, filters, pressure gauges and transfer lines. • Clean filters as required.	• Clean and recondition system.
Belts and Packing Glands		• Check belts for proper tension and damage. • Check packing glands for leakage and proper compressions.	
Air Leaks		• Check for leaks around access openings and flame scanner.	

Figure 9-7. Boiler Operations and Maintenance Requirements.

System	Daily Requirements	Weekly	Monthly	Annual
Air Leaks Waterside & Fireside Surfaces				• Clean surfaces according to manufacturer's recommendations.
Refractor on Fireside				• Repair refractor.
Electrical Systems				• Clean electrical terminals and replace defective parts.
Hydraulic & Pneumatic Valves				• Check all operations and repair all leakages.
Start-Up and Operation				• Check during start-up and operation.
Records	• Record type and amount of fuel used, exhaust gas temperature, and firing position and boiler room temperature.			

Figure 9-7. Boiler Operations and Maintenance Requirements (concluded)

OVENS (Monthly)

1. Inspect compartment for proper primary and secondary air conditions.
2. Regulate automatic pilot and safety valve for proper operation.
3. Check motor, belts, fans, on convection ovens.
4. Adjust thermostat for accurate calibration.
5. Check oven doors for (heat loss) tight fit.
6. Clean and adjust orifice and burner to rated BTU input.
7. Lubricate gas valves.
8. Adjust burner flame for proper gas/air mix.

PUMPS

Based on the pump manufacturer's recommendations, a PM Form of checks to be made should be incorporated. Checks should include:

1. Clean inside pump casing periodically and check impellor for wear or damage.
2. Check gland stuffing boxes and repack where necessary.
3. Check and adjust drives (as for fans).
4. Check nonreturn valves, pressure by-pass valves, etc., for correct and effective operation.

COMPRESSORS AND EVAPORATORS

1. *Weekly Checklist*
   a. Box temperature
   b. Thermostat setting
   c. Oil level of compressor (where appropriate)
   d. Flood back to compressor—no frost on compressor
   e. Operation of condenser and evaporator fans. Clean.
   f. Clean evaporator coils, pan and fans
   g. Leaks and oil spots
   h. Synchronization of timers (where applicable)
   i. Receiver temperature should be warm
   j. Short cycling
   k. Over heater strips, hardware

2. *Semi-Annual Checklist*
   a. Bank water level and immersion heater
   b. Leak test entire system
   c. Grease bearings on belt-driven fans
   d. Tighten all electrical terminals
   e. Check discharge pressure, receiver pressure, evaporator pressure, interstage pressure, and suction pressure as per manufacturer's recommendations
   f. Check expansion valve
   g. Check volts and amps of compressor and evaporator
   h. Noncondensibles in system
   i. Lowside pressure control setting. Cut in and cut out according to installation or condensed instruction.

Note: Do not make pressure adjustments without gauges installed, or without first checking recommended pressure setting in the manufacturer's instructions.

The ratio of brake horsepower consumed per ton of refrigerant output can vary considerably with the cleanliness of the condenser and evaporation. Table 9-1 indicates the measured variations of a nominal 15-ton capacity machine having a reciprocating compressor.

Table 9-1. The Effects of Poor Maintenance on the Efficiency of a Reciprocating Compressor, Nominal 15-Ton Capacity

Conditions	(1) °F	(2) °F	(3) Tons	(4) %	(5) HP	(6) HP/T	(7) %
Normal	45	105	17	—	15.9	0.93	—
Dirty Condenser	45	115	15.6	8.2	17.5	1.12	20
Dirty Evaporator	35	105	13.8	18.9	15.3	1.10	18
Dirty Condenser and Evaporator	35	115	12.7	25.4	16.4	1.29	39

(1) Suction Temp, °F
(2) Condensing Temp, °F
(3) Tons of refrigerant
(4) Reduction in capacity %
(5) Brake horsepower
(6) Brake horsepower per Ton
(7) Percent increase in compressor bh per/ton

It can be seen that in the worst case, a reduction in capacity of some 25% occured with an increase of 39% in power requirement per ton of refrigerant.

REFRIGERATION MAINTENANCE

1. Manufacturer's specifications should be followed for selection of *all* lubricants and refrigerants.
2. Inspect and repair any damage to insulation on duct work and piping to avoid temperature loss and damage from condensation.
3. Check for plugged spray nozzles on condenser.
4. Check for dirt on fan blades or rotors causing an unbalanced condition and vibration. Do not paint fan blades.
5. Do not over-lubricate blower bearings. This will avoid oil or grease being thrown on blades and acting as catch agents for dust and dirt.
6. Check for wasted condenser and cooling water in terms of gallons per minute per ton of refrigeration.
7. Check controls on outdoor air sources, so that outside air supply is increased when sufficiently cool to replace refrigerated air.
8. Check for air leakage around doors and transoms through worn weather stripping.
9. Check for worn gaskets on refrigerator doors.
10. Check pump impellers and packings on circulating pumps.
11. Check for clean condensers to avoid poor heat transfer.
12. Check for excessive head pressure and proper suction pressure for longer life of compressor.
13. Check for possibilities of reclaiming condensing water where applicable.
14. Check defrosting cycles to avoid power loss from frost buildup.
15. Check for condition of compressor valves and pistons.
16. Check air cool condenser for fin damage and clean.
17. Seasonal Maintenance. Towards the end of the cooling season, a complete check should be made of air-condi-

tioning equipment while it is still performing. The following should be included:

a. Possible replacement of controls, belts, air filters, refrigerant filter driers, and insulation.

b. Check to see whether units are increasing in power consumption or cooling water requirements. Taking one unit at a time out of service, service it for idleness, drain water, back-off packing glands, drain oil, flush bearings, add new oil, and clean catch pans and tanks.

c. Check for worn parts and compression clearance.

The above work should be done regardless of how well the machinery has operated during the previous season.

Table 9-2 shows the measured effect of dirty evaporators and condensers on a nominal 520-ton absorption chiller.

Table 9-2. The Effects of Poor Maintenance on the Efficiency of an Absorption Chiller, 520-Ton Capacity

Condition	Chilled Water °F	Tower Water °F	Tons	Reduction in Capacity %	Steam lb/ton/H	Per Cent
Normal	44	85	520	—	18.7	—
Dirty Condenser	44	90	457	12	19.3	3
Dirty Evaporator	40	85	468	10	19.2	2.5
Dirty Condenser and Evaporator	40	90	396	23.8	20.1	7.5

A reduction in output of 23.8% occurs at the worst case with an increase in steam consumption of 7-5% per ton of refrigerant.

FANS

1. Wheel shaft bearings on belt-driven units of all types with prelubricated pillow blocks and grease fittings should be relubricated every three (3) years. For normal operating conditions, use a grease conforming to NLGI No. 2 consistency.

Motor bearings are prelubricated and should be relubricated every three (3) to five (5) years. Consult instructions on motor. Motors not having pipe plugs or grease fittings in bearing housing can be relubricated by removing end shields from motor.

2. Check belt tension every six (6) months. Belt should depress its width when pressed firmly inward at mid-way point between the pulleys. Too much tension will damage bearings; belt should be tight enough to prevent slippage. When replacing belt, replace motor sheave if "shoulder" is worn in groove. Do not replace with a larger diameter pulley as this will overload the motor.

3. Clean fan (or blower) blades and check for blade damage, which may cause out-of-balance running.

4. Check fan casing and duct connections for air leakage.

FILTERS, COILS, STRAINERS, DUCTS, AND REGISTERS

1. *Filters*—Manufacturer's recommendations regarding the method and interval of cleaning/replacement should be followed. The manually-serviced type air filter requires periodic cleaning or replacement. The usual indication that cleaning/replacement is required is either (a) a decrease in air flow through the filter (up to 10%), or (b) an increase in resistance across the filter (more than 100%).

   Large installations having a number of filters, can arrange a maintenance program of cleaning/replacement on a rotated basis at a regular interval. In certain large duct installations and central air-handling units, it is possible to install simple manometers to indicate the pressure differential across the filter.

   Self-cleaning filters and precipitators should also be examined periodically to observe expiration of the disposable media or accumulation of sludge into the collecting pan. Many manufacturers provide indicators for their equipment to show when servicing is required.

2. *Coils*—The efficient operation of both cooling and heating coils depends largely upon the cleanliness of the heat-

transfer surface. Finned tube surfaces require particular attention and can be cleaned with detergents and high-pressure water using portable units.

Spray coil units may require chemical treatment for the build-up of algae and slime deposited by cooling water. Chemical cleaning can be most effective, but caution must be exercised with the choice of chemicals on certain metal surfaces.

3. *Strainers*—Regular cleaning of strainer screens keeps pressure losses in liquid systems to a minimum, thus saving pumping energy. It may be possible to replace fine-mesh strainer baskets with large mesh, without endangering the operation of the system. This again will reduce the pressure loss in the system and save pumping energy.

4. *Ducts*—Periodic opening and cleaning of the inside of ducts, plenum chambers, air-handling units, etc., to remove residually deposited dust and particulate matter. This will assist in keeping down the duty of the air filter, and maximizing the period between air filter servicing.

5. *Registers*—Periodically check for accumulation of material or other foreign matter behind registers. Check also the register seal to the duct, to ensure that all the conditioned air louvres are in the direction required.

Adjustable registers should be checked for setting, as these are sometimes moved by accident or by unauthorized personnel.

ELECTRIC MOTORS

Inspection of electric motors will cover the following:

1. Check electric starter contactors, and loose wire connections.
2. Using a meter, check the starting load and running load against rated loads.
3. Adjust the belt tension to a slight slackness on the top side.
4. Align the belt to avoid damage to belts, bearings, and excessive electrical consumption.

5. Check bearings for wear, dust and dirt.
6. Check internal insulation to see that it is free of oil.
7. Check commutator slots and motor housing for dust and good air circulation.
8. Examine fusing and current limiting devices for protection while starting and then while running.
9. Check brushes for wear.

LEAKS - STEAM, WATER, AND AIR

The importance of leakage cannot be understated. If a plant has many leaks, this may be indicative of a low standard of operation involving the loss not only of steam, but also water, condensate, compressed air, etc.

If, for example, a valve spindle is worn, or badly packed, giving a clearance of 0.010 inch between the spindle, for a spindle of ¾-inch diameter, the area of leakage will be equal to a 3/32-inch diameter hole. Table 9-3 illustrates fluid loss through small holes:

Table 9-3. Fluid Loss Through Small Holes

Diameter of Hole	Steam — lb/hour 100 psig	300 psig	Water -- gals/hour 20 psig	100 psig	Air SCFM 80 psig
1/16"	14	33	20	45	4
1/8"	56	132	80	180	16
3/16"	126	297	180	405	36
1/4"	224	528	320	720	64

Although the plant may not be in full production for every hour of the entire year (i.e., 8760 hours), the boiler plant water systems and compressed air could be operable. Losses through leakage are usually, therefore, of a continuous nature.

THERMAL INSULATION

Whatever the pipework system, there is one fundamental–it should be adequately insulated. Table 9-4 gives a guide to the degree of insulation required. Obviously there are a number of

Table 9-4. Pipe Heat Losses

Pipe Dia Inches	Surface Temp°F	Insulation Thickness Inches	Heat Loss (BTU/Ft/Hr)		Insulation Efficiency
			Uninsulated	Insulated	
4	200	1½	300	70	76.7
	300	2	800	120	85.0
	400	2½	1500	150	90.0
6	200	1½	425	95	78.7
	300	2	1300	180	85.8
	400	2½	2000	195	90.25
8	200	1½	550	115	79.1
	300	2	1500	200	86.7
	400	2½	2750	250	91.0

types of insulating materials with different properties and at different costs, each one of which will give a variancy return on capital. Table 9-4 is based on a good asbestos or magnesia insulation, but most manufacturers have cataloged data indicating various benefits and savings that can be achieved with their particular product.

STEAM TRAPS

The method of removing condensate is through steam trapping equipment. Most plants will have effective trapping systems. Others may have problems with both the type of traps and the effectiveness of the system.

The problems can vary from the wrong type of trap being installed, to air locking, or steam locking. A well-maintained trap system can be a great steam saver. A bad system can be a notorious steam waster, particularly where traps have to be bypassed or are leaking.

Therefore, the key to efficient trapping of most systems is good installation and maintenance. To facilitate the condensate removal, the pipes should slope in the direction of steam flow. This has two obvious advantages in relationship to the removal of condensate; one is the action of gravity, and the other the pushing action of the steam flow. Under these circumstances the strategic siting of the traps and drainage points is greatly simplified.

One common fault that often occurs at the outset is installing the wrong size traps. Traps are very often ordered by the size of the pipe connection. Unfortunately the pipe connection size has nothing whatsoever to do with the capacity of the trap. The discharge capacity of the trap depends upon the area of the valve, the pressure drop across it, and the temperature of the condensate.

It is therefore worth recapping exactly what a steam trap is. It is a device that distinguishes between steam and water and automatically opens a valve to allow the water to pass through but not the steam. There are numerous types of traps with various characteristics. Even within the same category of traps, e.g., ball floats or thermoexpansion traps, there are numerous designs, and the following guide is given for selection purposes:

1. Where a small amount of condensate is to be removed an expansion or thermostatic trap is preferred.
2. Where intermittent discharge is acceptable and air is not a large problem, inverted bucket traps will adequately suffice.
3. Where condensate must be continuously removed at steam temperatures, float traps must be used.
4. When large amounts of condensate have to be removed, relay traps must be used. However, this type of steam trap is unlikely to be required for use in the food industry.

To insure that a steam trap is not stuck open, a weekly inspection should be made and corrective action taken. Steam trap testing can utilize several methods to insure proper operation:

- Install heat sensing tape on trap discharge The color indicates proper operation.
- Place a screw driver or more sophisticated acoustical instrument to the ear lobe with the other end on the trap. If the trap is a bucket-type, listen for the click of the trap operating.

CONTROL DEVICES

The functional operation of control equipment is of no use unless the equipment operates correctly at the required set point. Periodic checking and recalibration of all control equipment is an essential aspect of energy conservation.

1. *Thermostats*—In many cases, thermostats can be checked with a mercury-in-glass thermometer, and calibration adjustments can be made. Temperature differential for a signal is not usually adjustable. If it is found that the differential is too great, then usually it is necessary to replace the unit. Checks should be made at both maximum and minimum set positions.

2. *Humidistats*—These can be checked with a wet and dry bulb thermometer. Most units can be easily recalibrated, but operating differentials across a set point cannot usually be adjusted.

3. *Control Valves*—These should be checked and adjusted for operation by monitoring the actuating signal with a known standard,* or by using an auxiliary signal for an alternative corrected source.

ADJUSTMENTS

1. *Actuators*—These should be checked for operation (and repetition of operation) from a signal. Length of stroke, or angle of arc should be checked to ensure full operational movement.

2. *Linkages*—Check for ease of motion; lubricate fulcrum and check for heat. Check locking devices on adjustable linkages and make sure that they are in the original position determined during the testing and balancing of the system.

3. *Motor Drives*—These are used for the control of some

---

* Direct acting valves—mercury/glass thermometer
  Pneumatic valves—pressure gauge
  Electrically operated valves—ammeter/voltmeter.

types of valves, dampers, etc. Check for length of stroke/ angle of arc, security of fixing, and adjust where required,

4. *Manual Dampers*—Check that these are set at positions determined during the testing and balancing of the system. Check for leakages around the spindle, and check that the quadrant permits full open/close operation of the blade. Check that the blades give a tight shut-off.

5. *Registers*—Check that these are set to discharge the air in the direction required. Make sure that short-circuiting of delivered conditioned air into the return air system *does not* occur.

10

Self-Evaluation Checklists

INTRODUCTION

The self-evaluating checklists are to be used to:

1. Determine the major factors of energy consumption in the federal facility and determine factors contributing to the overall energy usage in the specific area.

2. Discover transferable techniques for saving energy.

3. Provide guidance to federal facility managers to pinpoint modifications in building systems and operational practices that would result in reducing energy consumption.

4. Identify areas where additional information would be helpful and constructive suggestions welcome.

The initiative and the responsibility of corrective actions remain with the individual manager. To aid in this analysis, these self-evaluating checklists have been developed. They provide the manager with an indication of the factors of thermal performance which require correction.

The checklists consist of separate sections or areas of evaluation. A relative numerical value has been assigned to those specific conditions that effect the energy loss in these areas. Additional instructions are also provided in the self-evaluating checklists to assist in completion of the form and provide consistency of results between federal facilities. When completing the forms using these instructions and computing a resultant overall score for each section, both strong and weak areas become apparent. This scoring method is valid for each of the sections as well as for each item within a section.

The purpose of these checklists is to assist in dealing with the "How" of starting an energy management program. In this handbook space does not permit listing the recommendations of possible remedies for the twenty evaluation sections. Each manager must determine the best use of budgeted expenditures for reducing energy consumption.

INSTRUCTIONS FOR SELF-EVALUATION CHECKLISTS

To demonstrate the use of self-evaluating checklists, an example is presented as follows:

Seven windows are used to demonstrate the typical checklist shown in the following example. Note that each window condition is assigned a value if the condition applies. The overall rating for the windows listed is 51 percent. This rating scale of 51 percent indicates that corrective action is required in this area since its rating is only half of the maximum score of 100 allowable.

Although the example covers only seven windows, a typical building evaluation will include hundreds of windows. Each form provides for the listing of 25 windows. A sufficient number of forms should be used to list each window as an individual item.

For record-keeping purposes, it is suggested that each window be assigned an "address" which will serve to positively identify that opening for all references regarding that window. Architectural building exterior elevation drawings will be useful as a means of tabulating and recording work on windows.

There are twenty categories or evaluation sections, as follows:
1. Window
2. Door
3. Ceiling
4. Wall
5. Roof

NO.	LOCATION	RATING VALUE MAX. = 10	Storms	Solar Protection	Tight Fit	Minor Infiltration	Major Infiltration	Cannot Be Opened	Can Be Opened	Weather Stripped							TOTAL POINTS
			2	2	2	1	0	3	0	1						
1	Bldg. 4, Room 401		2			1			0	1							4
2	Bldg. 4, Room 402			2	2			3		1							8
3	Bldg. 4, Room 609					1			0	1							2
4	Bldg. 4, Room 102		2	2	2			3		1							10
5	Bldg. 4, Room 104, W1		2			1			0	1							4
6	Bldg. 4, Room 104, W2		2				0		0	1							3
7	Bldg. 4, Room 104, W3		2		2				0	1							5
25	GRAND TOTAL																36

EVALUATOR A. AUD
DATE 5/10/74
UNIT NAME Anywhere
SHEET NO. 1

WINDOW CONDITIONS

$$\text{RATING SCORE} = 100 \times \frac{36}{(7)(10)} = 51\%$$

6. Storage Area
7. Shipping and Receiving
8. Illumination
9. Food Area
10. Heat Generation
11. Heat Distribution
12. Cooling Generation
13. Cooling Distribution
14. Electrical Power Distribution
15. Hot Water Service
16. Laundry
17. Compressed Air
18. Water
19. Process Heating
20. Transportation

Completion of these 20 forms by the manager and/or his staff will provide the current status of energy consumption in those areas identified as needing the most attention. Recommendations to improve any faulty conditions should be evaluated using an energy savings cost analysis for these 20 parameters to ensure that the greatest energy savings per dollar are attained.

Each of the 20 sections will be evaluated on a separate rating schedule. There are three parts to each section:

1. Recommendations for improvements (not included in this text)
2. Instructions for evaluating ratings
3. Checklist

Specific conditions in each category are determined by completing the corresponding checklist. Each item being evaluated is identified and located in the appropriate space on the form. An example would be Building 4, Room 406, Window 1.

Each of the specific conditions listed on the checklist is evaluated for each item. The instruction sheet provides guidance in properly identifying correct conditions. The assigned value for each existing condition should be listed in the proper column to credit the item being evaluated. Total points for each item are determined and this total listed in the item total points column.

Each form will accommodate 25 items of similar nature, such as 25 windows or 25 doors. As many forms as are necessary to list all similar items on an individual line should be used. The total points for each section are determined by adding all item total points for that section.

Using the following scoring formula, the rating score of each of the 20 sections should be individually calculated.

$$\text{Rating Score} = \frac{(100) \times (\text{Point Total for Section})}{(\text{No. of Items}) \times (\text{Maximum Rating Value in Section})}$$

This rating score is then applied to the following table which indicates the urgency of corrective action.

Range of Rating Score	Action Required
0 – 20	Immediate Corrective Action Required
20 – 40	Urgent Corrective Action Required
40 – 60	Corrective Action Required
60 – 80	Evaluation for Potential Improvement Required
80 – 100	No Corrective Action Required

Recommendation sheets are included in the government guide that list several methods to improve the score in each section. These recommendations are general in nature. The recommendation that is prevalent in all sections is: Education and training of personnel to reduce energy consumption. It is critical that managers realize the importance of the individual's role in a personal commitment to energy conservation. Employee awareness of energy consumption and its reduction should be given high priority when establishing energy conservation policies and practices.

In the following pages are checklists and instruction for their use for each of the 20 sections.

WINDOW RATING INSTRUCTIONS

2 points   if the window has storm windows adequate for cold weather protection. The storm windows must fit tightly and block the wind from entering around the window.

2 points   if the window has protection from the direct sun during warm weather. Solar protection can be part of the building design such as overhang, awnings or physical shields. Protection can also be tinted or reflective film applied to the windows, double-glazed windows, solar screening or trees blocking out direct sunlight.

2 points   for a tight fitting window. A window is tight fitting if the infiltration will not be detected around the window during a windy day. The window must fit well and all caulking must be in place. Weatherstripping will contribute to a tight fit.

1 point   if the wind has some infiltration around the window. The window should fit fairly well and not be loose and rattle.

0 points   if infiltration can be felt to a large degree. The window is loose in the frame and caulking is missing or in poor condition.

3 points   if the window is designed so physically it cannot be opened.

0 points   if it can be opened. If it can be opened, it will be opened to "regulate" room temperature.

1 point   if window is weatherstripped all around and the weatherstripping is in good condition.

SELF-EVALUATING CHECKLIST FOR WINDOWS

EVALUATOR _____

DATE _____

UNIT NAME _____

SHEET NO. _____

NO.	LOCATION	RATING VALUE MAX. = 10	Storms	Solar Protection	Tight Fit	Minor Infiltration	Major Infiltration	Cannot Be Opened	Can Be Opened	Weather Stripped								TOTAL POINTS
			2	2	2	1	0	3	0	1								
1																		

DOOR RATING INSTRUCTIONS

This section applies to all doors that open to the outside and all doors that open to an unconditioned space such as warehouses and storerooms.

2 points   if door is part of an air-lock system.
1 point   if door has a closer which may be either spring, air or hydraulic.
1 point   if door closer does not have a hold-open feature.
0 points   if door closer has a hold-open feature.
2 points   if door fits snugly into the door frame with no loose condition and where no infiltration exists around the edges.
1 point   if door is an average fit and can be slightly rattled in the frame and has a slight infiltration around the edges.
0 points   if door is loose in the frame and infiltration exists.
2 points   if weatherstripping exists on all four edges and is in good condition. (Thresholds with elastic or fiber to close the space, and astragals on double doors are considered weatherstripping.)
1 point   if weatherstripping exists on jambs and head only.
0 points   if no weatherstripping exists or if it exists and is in poor condition.
1 point   if door is protected from outside wind. This can be building design, wind screen or shrubbery.

SELF-EVALUATING CHECKLIST FOR EXTERIOR DOORS

EVALUATOR			DOOR CONDITIONS																	
DATE			Air Lock	Door Has Closer	Closer Has No Hold-Open	Closer Has a Hold-Open	Snug Fit	Average Fit	Loose Fit	Weather Strip 4 Edges	Weatherstrip Jamb Head	No Weatherstrip	Wind Screens or Other							TOTAL POINTS
UNIT NAME																				
SHEET NO.																				
NO.	LOCATION	RATING VALUE MAX. = 10	2	1	1	0	2	1	0	2	1	0	1							
1																				

CEILING RATING INSTRUCTIONS

1 point   if a drop ceiling exists.
1 point   if insulation exists above ceiling on top floor below roof or mechanical space.
1 point   if space above drop ceiling is mechanically vented. Natural draft is not considered mechanical venting.
2 points   if all panels are in place and in good condition, no broken or missing panels are present.
1 point   if panels are broken or in poor condition.
0 points   if panels are missing or removed and out of place.

			RATING	Drop Ceiling	Insulated Drop Ceiling	Insulated Reg. Ceiling	Space Not Mech. Vented	All Panels in Place	Panels Broken	Panels Missing										TOTAL POINTS
			CEILING CONDITIONS																	
NO.	LOCATION		VALUE MAX. = 6	1	1	1	1	2	1	0										
1																				

EVALUATOR _____
DATE _____
UNIT NAME _____
SHEET NO. _____

WALL RATING INSTRUCTIONS

3 points if wall is designed to resist outside temperature differential. Insulation is present to substantially change heat transfer time.

0 points if wall is merely a physical separation without adequate insulating qualities.

2 points if outside wall surface has solar protection such as light finish, is heavily shaded or has physical sun screens.

2 points if surfaces of walls are in good repair and not damaged.

1 point if inside is in average condition with a few small cracks in the surface and smaller plaster sections missing.

0 points if wall has openings to unconditioned space; i.e., plumbing or duct openings not closed.

SELF-EVALUATING CHECKLIST FOR EXTERIOR WALLS

EVALUATOR _____
DATE _____
UNIT NAME _____
SHEET NO. _____

			RATING	Insulated	Not Insulated	Solar Protection	Watertight	Cracked or Broken	Open to Noncondition Sp.										TOTAL POINTS
			WALL CONDITIONS																
NO.	LOCATION		VALUE MAX. = 7	3	0	2	2	1	0										
1																			

ROOF RATING INSTRUCTIONS

2 points   if roof insulation is in dry condition.
0 points   if roof insulation is in poor condition, wet, aged, brittle, cracked, etc., or if no insulation exists.
1 point   if roof has a reflective surface; this may be the type of material used or the color and condition of surface (gravel, etc.).
1 point   if mechanical ventilation exists between roof and ceiling below. This should be properly sized so adequate air flow exists.
2 points   if no leaks exist in the roof.
1 point   if minor leaks exist.
0 points   if there are many leaks.

SELF-EVALUATING CHECKLIST FOR ROOFS

EVALUATOR _____

DATE _____

UNIT NAME _____

SHEET NO. _____

NO.	LOCATION	RATING VALUE MAX. = 6	Dry Insulation	Wet Insulation	Reflective Surface	Ventilation Under Roof	No Leaks	Small Leaks	Many Leaks											TOTAL POINTS
			2	0	1	1	2	1	0											
1																				

STORAGE AREA RATING INSTRUCTIONS

1 point   if area is not temperature controlled.
1 point   if the doors are kept closed.
2 points   if there are no windows in the area.
1 point   if one window is in the area.
0 points   if two or more windows are in the area.
2 points   if area is used as it was designed.
0 points   if area is used for storage but designed for other usage.

SELF-EVALUATING CHECKLIST FOR STORAGE AREAS

EVALUATOR _____

DATE _____

UNIT NAME _____

SHEET NO. _____

NO.	LOCATION	RATING VALUE MAX. = 6	Not Conditioned	Door Closed	No Windows	One Window	Two or More Windows	Used as Designed	Not Used as Designed									TOTAL POINTS
			1	1	2	1	0	2	0									
1																		

SHIPPING AND RECEIVING AREA RATING INSTRUCTIONS

3 points   if the shipping and receiving area is well protected from outside temperature.

1 point   if the shipping and receiving area is reasonably protected from outside air entry.

0 points   if the shipping and receiving area has no protection from the ambient. This would be an open area directly exposed to the outside conditions.

1 point   if individual truck stalls exist so the unused areas can be closed.

0 points   if one large area exists and the entire dock must be exposed if a single truck is loaded or unloaded.

1 point   if the doors are closed when not in use.

0 points   if the doors are left open as a matter of convenience.

1 point   if the area does not receive conditioned air.

0 points   if the area receives conditioned air.

SELF-EVALUATING CHECKLIST FOR SHIPPING AND RECEIVING AREAS

			SHIPPING AND RECEIVING CONDITIONS																
EVALUATOR _____																			
DATE _____																			
UNIT NAME _____			Weather Prot. Good	Weather Prot. Average	Weather Prot. Poor	Individual Stalls	One Large Area	Doors Closed	Doors Opened	Not Temp. Cond.	Temp. Cond.								TOTAL POINTS
SHEET NO. _____																			
NO.	LOCATION	RATING VALUE MAX. = 6	3	1	0	1	0	1	0	1	0								
1																			

ILLUMINATION RATING INSTRUCTIONS

1 point   if extensive decorative lighting has been eliminated where used for reasons of appearances (not security, walkway lighting and other necessities).

1 point   if lighting has been arranged to illuminate only the work area.

0 points   if lighting has been designed to illuminate the entire room to a working level.

2 points   if light fixture diffuser is clean and clear.

1 point   if diffuser is slightly yellowed or dirty.

0 points   if diffuser is noticeably yellowed or dust is visible. This restriction can amount to 10% or more of the light flux being transmitted.

2 points   if fixture internal reflective surface is in good condition (the paint is reflective and clean).

1 point   if the fixture internal reflective surface gives dirt indication on clean white cloth.

0 points   if the reflective surface is yellowed and dull.

1 point   if fluorescent lights are used for all illumination.

0 points   if incandescent lights are used.
1 point   if lights are properly vented so the heat can escape to ceiling space, providing that ceiling space is ventilated to prevent heat build-up.
1 point   if lights are turned off when area is not occupied.
1 point   if illumination level is adequate for designed usage.
0 points   if area is "over illuminated" for designed usage.*
0 points   if two or more lamps have blackened ends or are glowing without lighting.

SELF-EVALUATING CHECKLIST FOR ILLUMINATION

EVALUATOR _____
DATE _____
UNIT NAME _____
SHEET NO. _____

NO.	LOCATION	RATING VALUE MAX. = 10	No Decorative Ltg.	Light Work Area	Light Entire Room	Diffusers Good	Diffusers Average	Diffusers Poor	Reflection Good	Reflection Average	Reflection Poor	Flourescent Lights	Incandescent Lights	Lights Vented	Lights Turned Off	Illumination Adeq.	Excessive Illumination			TOTAL POINTS
			1	1	0	2	1	0	2	1	0	1	0	1	1	1	0			
1																				

*Note: Momentarily disconnect lamps until level is reached which is adequate for the intended function. The following light meter readings will assist in determination of average adequate light levels. These are below Illumination Engineering Society recommendations in some instances. Absence of reflected glare is mandatory for reading tasks requiring careful fixture placement.

Corridors, lobbies	—10-15 footcandles average
Typing areas	—50 footcandles in area of work, 20 elsewhere
Storerooms	—5 footcandles
Prolonged reading task areas	—50 footcandles
Kitchens	—50 footcandles in areas of work, 20 elsewhere
Laboratories	—50 footcandles in areas of work
Toilet rooms	—20 footcandles at mirrors

Federal Energy Administration
Recommended Maximum Lighting Levels

Task or area	Footcandle levels	How measured
Hallways or corridors	10±5	. . . . . .Measured average, minimum 1 footcandle.
Work and circulation areas surrounding work stations	30±5	. . . . . .Measured average.
Normal office work, such as reading and writing (on task only), store shelves, and general display areas	50±10	. . . . . .Measured at work station.

Task or area	Footcandle levels	How measured

Prolonged office work
which is somewhat difficult
visually (on task only) . . . . 75±15 . . . . . .Measured at work station.
Prolonged office work
which is visually difficult
and critical in nature
(on task only) . . . . .100±20 . . . . . .Measured at work station.

FOOD AREA RATING INSTRUCTIONS

2 points   if the food preparation equipment is only energized when actually needed. This includes, but is not limited to, ovens, warmers, steam tables, delivery equipment and coffee urns.
0 points   if equipment is turned on and left on all day.
1 point   if refrigerator and freezer doors are kept tightly closed.
0 points   if refrigerator and freezer doors can be left ajar.
1 point   if faucets and valves are in good condition and not leaking.
0 points   if faucets and valves are leaking. Leaks may be external or internal in the system.
3 points   if doors between kitchen area and other areas are kept closed.
2 points   if adequate vent hoods are used over heat-producing equipment.
1 point   if some vent hoods are used over heat-producing equipment.
0 points   if no or inadequate vent hoods are used.
1 point   if ventilation air supply is adequate to remove most of the heat produced by the kitchen equipment.
2 points   if refrigerator equipment is in good repair, seals are good, condenser is clean, air passage over condenser is clear.
1 point   if refrigeration equipment is in average condition, dust and dirt exist on condensers but the air flow is not restricted, door gaskets seal all around although they may have lost some resiliency.
0 points   if refrigeration equipment is in poor condition, a large collection of dust and dirt on the condenser or the fins may be bent to restrict air flow, door gaskets do not seal all around, are brittle, broken or missing.
3 points   if heat-recovery systems are utilized. These can be applied to the exhaust air, the hot waste water or on the refrigeration equipment.

SELF-EVALUATING CHECKLIST FOR FOOD AREA

EVALUATOR_____
DATE_____
UNIT_____
NAME
SHEET NO. _____

NO.	LOCATION	RATING VALUE MAX. = 15	Equipment Turned Off	Equipment Left On	Refrig. Doors Closed	Refrig. Doors Ajar	Faucets Not Leaking	Faucets Leaking	Access Doors Closed	Good Vent Hoods	Average Vent Hood	Poor Vent Hood	Adequate Ventilation	Refrig. Equip. Good	Refrig. Equip. Average	Refrig. Equip. Poor	Heat Recovery System				TOTAL POINTS
			2	0	1	0	1	0	3	2	1	0	1	2	1	0	3				
1																					

HEATING SYSTEM (GENERATION) RATING INSTRUCTIONS

2 points   if the insulation is in good condition with no broken or missing sections. The insulation must not be wet, crumbly or cracked.

1 point   if insulation is in average condition with small sections broken or missing. The insulation must not be wet or crumbly.

0 points   if insulation is in poor condition with sections missing, broken, wet, crumbly or cracked.

2 points   if flanges, valves and regulators are insulated with removable lagging.

2 points   if the steam system has no leaks.

1 point   if the steam system has minor leaks around valve packing, shaft seals, etc.

0 point   if the steam system has many leaks, valves, regulators and traps have dripping leaks, steam plumes, etc.

1 point   if boiler combustion controls are automatic.

1 point   if definite standard operating procedures are used. These should be written and posted near the boiler control panel.

1 point   if each boiler has an individual steam flow meter.

1 point   if each boiler has an individual make-up water meter.

1 point   if each boiler has an individual fuel flow meter.

1 point   if a definite preventive maintenance schedule is followed.

0 points   if equipment is maintained or repaired only when it breaks down.

3 points   if an energy recovery system is used. This may be a heat exchanger of water to water, an air wheel or any of several types in common use.

2 points   if heat generation is controlled by a system using an economizer system by comparing inside and outside temperature.

SELF-EVALUATING CHECKLIST FOR HEAT GENERATION

			HEAT GENERATION CONDITIONS															
EVALUATOR _____			Insulation Good	Insulation Average	Insulation Poor	Flanges Insulated	No Leaks	Some Leaks	Many Leaks	Auto Controls	Standard Op. Procedure	Steam Meter	Fuel Meter	Make-Up Water Meter	Preventive Maintenance	Fix as Required Schedule	Energy Recovery	Economizer Controls	TOTAL POINTS
DATE _____ UNIT NAME _____ SHEET NO. _____																			
NO.	LOCATION	RATING VALUE MAX. = 17	2	1	0	2	2	1	0	1	1	1	1	1	1	0	3	2	
1																			

HEATING SYSTEM (DISTRIBUTION) RATING INSTRUCTIONS

2 points   if insulation is in good condition with no broken or missing sections. The insulation must not be wet, crumbly or cracked.

1 point   if insulation is in average condition with small sections broken or missing. The insulation must not be wet, crumbly or cracked.

0 points   if insulation is in poor condition with sections missing, broken, wet, crumbly or cracked.

2 points   if flanges, valves and regulators are insulated with removable lagging.

2 points   if the steam system has no leaks.

1 point   if the steam system has minor leaks around valve packing, shaft seals, etc.

0 points   if the steam system has many leaks, valves, regulators and traps have dripping leaks, steam plumes, etc.

2 points   if the control system to each area is adequate. The control system shall maintain the temperature in each room close to the thermostat setting.

1 point   if the control system to each area is only a general control without the ability to control each room.

0 points   if the control system has little or no control over the area temperature. Also included here is a control system that allows the heating and cooling systems to oppose each other in the same general area.

1 point   if definite standard operating procedures are used. These should be written and posted.

1 point   if a definite preventive maintenance schedule is followed.

0 points   if equipment is maintained or repaired only when it breaks down.

1 point   if the area is conditioned only when occupied. This will apply especially to auditoriums, work rooms, hobby shops, TV rooms, etc.

0 points   if the area is conditioned all the time regardless of occupancy.

2 points   if the zone control is good and certain areas can be secured when not in use or require less temperature conditioning.

1 point   if the zone control only allows general areas to be secured when conditions dictate.

0 points   if zone control cannot be secured without securing a large general area.

SELF-EVALUATING CHECKLIST FOR HEAT DISTRIBUTION

EVALUATOR _____

DATE _____

UNIT NAME _____

SHEET NO. _____

			HEAT DISTRIBUTION CONDITIONS																	
NO.	LOCATION	RATING VALUE MAX. = 13	Insulation Good	Insulation Average	Insulation Poor	Flanges Insulated	No Leaks	Some Leaks	Many Leaks	Control Good	Control Average	Control Poor	Standard Op. Procedure	Preventive Maintenance	Fix as Required	Condition as Required	Constant Conditioning	Zone Control Good	Zone Control Average	Zone Control Poor	TOTAL POINTS
			2	1	0	2	2	1	0	2	1	0	1	1	0	1	0	2	1	0	
1																					

COOLING SYSTEM (GENERATION) RATING INSTRUCTIONS

2 points   if the insulation is in good condition with no broken or missing sections. The insulation must not be wet, crumbly or cracked. Closed cell insulation will be considered average condition because of deterioration that occurs in this type of material.

1 point	if insulation is in average condition with small sections broken or missing. The insulation must not be wet or crumbly. The outside shell of open cell insulation must be intact with only minor breaks.
0 points	if insulation is in poor condition with sections missing, broken, wet, crumbly or cracked.
1 point	if flanges and valves are insulated.
1 point	if definite standard operating procedures are used. These should be written and posted near the control panel.
1 point	if unit has an individual watt-hour meter so the real-time power consumption can be determined.
1 point	if a definite preventive maintenance schedule is followed.
0 points	if equipment is maintained or repaired only when it breaks down.
3 points	if an energy recovery system is used. This may be a heat exchanger of water to water, an air wheel or any of several types in common use.
2 points	if outside air is utilized to help condition areas that require cooling even on cold days.
1 point	if the fresh air ratio is regulated by comparing inside requirements with outside temperatures.

SELF-EVALUATING CHECKLIST FOR COOLING SYSTEM GENERATION

			Insulation Good	Insulation Average	Insulation Poor	Flanges Insulated	Standard Op. Procedure	Ind. Power Meter	Preventive Maintenance	Fix as Required	Energy Recovery	Outside Air Used	Req. Fresh Air						TOTAL POINTS
NO.	LOCATION	RATING VALUE MAX. = 12	2	1	0	1	1	1	1	0	3	2	1						
1																			

EVALUATOR_____
DATE_____
UNIT NAME _____
SHEET NO. _____

COOLING GENERATION CONDITIONS

COOLING SYSTEM (DISTRIBUTION) RATING INSTRUCTIONS

2 points	if the insulation is in good condition with no broken or missing sections. The insulation must not be wet, crumbly or cracked. "Closed cell" insulation will be considered average condition because of deterioration that occurs in this type of material.
1 point	if insulation is in average condition with small sections broken or missing. The insulation must not be wet, crumbly. The outside shell of "open cell" insulation must be intact with only minor breaks.
0 points	if insulation is in poor condition with sections missing, broken, wet, crumbly or cracked.
1 point	if flanges and valves are insulated.
1 point	if definite standard operating procedures are used. These should be written and posted near the control panel.

2 points if the control system to each area is adequate. The control system shall maintain the temperature in each room close to the thermostat setting.

1 point if the control system to each area is only a general control without the ability to control each room.

0 points if the control system has little or no control over the area temperature. Also included here is a control system that allows the heating and cooling systems to oppose each other in the same general areas.

1 point if a definite preventive maintenance schedule is followed.

0 points if equipment is maintained or repaired only when it breaks down.

1 point if the area is conditioned only when occupied. This will apply especially to auditoriums, work rooms, hobby shops, TV rooms, etc.

0 points if the area is conditioned all the time regardless of occupancy.

2 points if the zone control is good and certain areas can be secured when not in use or require less temperature conditioning.

1 point if the zone control only allows general areas to be secured when conditions dictate.

0 points if zone control cannot be secured without securing a large general area.

SELF-EVALUATING CHECKLIST FOR COOLING DISTRIBUTION

EVALUATOR _____

DATE _____

UNIT NAME _____

SHEET NO. _____

NO.	LOCATION	RATING VALUE MAX. = 11	Insulation Good	Insulation Average	Insulation Poor	Flange Insulated	Standard Op. Procedure	Controls Good	Controls Average	Controls Poor	Preventive Maintenance	Fix as Required	Condition as Required	Constant Conditioning	Zone Control Good	Zone Control Average	Zone Control Poor					TOTAL POINTS
			2	1	0	2	1	2	1	0	1	0	1	0	2	1	0					
1																						

ELECTRICAL POWER DISTRIBUTION RATING INSTRUCTIONS

2 points for operation of a recording ammeter.

1 point for hourly electrical usage pattern of building being determined.

1 point for study of electrical requirements with the Power Company staff.

1 point for installation of a power peak warning system.

1 point for analysis to eliminate power peak demands.

1 point if a definite standard operating procedure is used. This shall be written and posted near the control panel.

1 point if definite preventive maintenance schedule is followed.

0 points if equipment is maintained or repaired only when it breaks down.

2 points for overall system Power Factor of 90 percent or above at main service.

SELF-EVALUATING CHECKLIST FOR ELECTRICAL POWER DISTRIBUTION

EVALUATOR___ DATE___ UNIT NAME ___ SHEET NO. ___			POWER DISTRIBUTION CONDITIONS																	
NO.	LOCATION	RATING VALUE MAX. = 10	Recording Meter	Usage Pattern	Power Co. Coord.	Power Peak Warning	Power Demand Limited	Standard Op. Procedure	Preventive Maintenance	Fix as Required	90% Power Factor									TOTAL POINTS
			2	1	1	1	1	1	1	0	2									
1																				

HOT WATER SERVICE RATING INSTRUCTIONS

2 points — if the insulation is in good condition with no broken or missing sections. The insulation must not be wet, crumbly or cracked.

1 point — if insulation is in average condition with small sections broken or missing. The insulation must not be wet or crumbly.

0 points — if insulation is in poor condition with sections missing, broken, wet, crumbly or cracked.

1 point — if faucets and valves are in good repair.

0 points — if faucets and valves leak externally or internally.

1 point — if definite standard operating procedures are used. These should be written and posted.

1 point — if a definite preventive maintenance schedule is followed.

0 points — if equipment is maintained or repaired only when it breaks down.

SELF-EVALUATING CHECKLIST FOR HOT WATER SERVICE

EVALUATOR___ DATE___ UNIT NAME ___ SHEET NO. ___			HOT WATER SERVICE CONDITIONS																
NO.	LOCATION	RATING VALUE MAX. = 5	Insulation Good	Insulation Average	Insulation Poor	No Faucet Leaks	Faucet Leaks	Standard Op. Procedure	Preventive Maintenance	Fix as Required									TOTAL POINTS
			2	1	0	1	0	1	1	0									
1																			

LAUNDRY RATING INSTRUCTIONS

2 points if overall equipment is in good condition. This means all equipment is operating per manufacturers' specifications. There are no leaks; gaskets and seals are all functioning properly, nothing is "jury rigged" to enable it to work, and equipment is used for its designed function, etc.

1 point if overall equipment is in average shape. Equipment condition will deteriorate over time due to normal usage. If equipment has been in use for a few years it should be placed in this category.

0 points if equipment is in poor condition. This includes leaks, malfunctioning equipment, improperly adjusted components, bypassing manufacturers' operational procedures, etc.

1 point if faucets, valves and traps are in good condition. Faucets should not leak, valves should seal tight and traps cannot have any blow by.

3 points if energy recovery systems are used. These can be any of several systems on the market today.

1 point if the laundry hot water generator is secured during laundry off periods such as evenings and weekends. An analysis should be made of each hot water system to determine the recovery time to ensure hot water is available when required.

2 points if the insulation is in good condition with no broken or missing sections. The insulation must not be wet, crumbly or cracked.

1 point if insulation is in average condition with small sections broken or missing. The insulation must not be wet, or crumbly.

0 points if insulation is in poor condition with sections missing, broken, wet, crumbly or cracked.

2 points if flanges, valves and regulators are insulated with removable lagging.

1 point if definite standard operating procedures are used. These should be written and posted.

1 point if a definite preventive maintenance schedule is followed.

0 points if equipment is maintained or repaired only when it breaks down.

SELF-EVALUATING CHECKLIST FOR LAUNDRY

NO.	LOCATION	RATING VALUE MAX. = 13	Equip. Condition Good	Equip. Condition Average	Equip. Condition Poor	No Leaks	Energy Recovery System	Hot Water Gen. Secured	Insulation Good	Insulation Average	Insulation Poor	Flanges Insulated	Standard Op. Procedure	Preventive Maintenance	Fix as Required				TOTAL POINTS
			2	1	0	1	3	1	2	1	0	2	1	1	0				
1																			

EVALUATOR_____

DATE_____

UNIT NAME_____

SHEET NO. _____

COMPRESSED AIR SERVICE RATING INSTRUCTIONS

1 point	if outlets and valves are in good repair.
0 points	if outlets and valves leak externally or internally.
1 point	if compressors are properly sized to shave peak demands.
1 point	if additional compressors are brought on line as demand requires and not run continuously.
1 point	if definite standard operating procedures are used. These should be written and posted.
1 point	if a definite preventive maintenance schedule is followed.
0 points	if equipment is maintained or repaired only when it breaks down.

SELF-EVALUATING CHECKLIST FOR COMPRESSED AIR

EVALUATOR _____

DATE _____

UNIT NAME _____

SHEET NO. _____

COMPRESSED AIR CONDITIONS

NO.	LOCATION	RATING VALUE MAX. = 5	No Outlet Leaks	Outlet Leaks	Compressors Sized	Compressors on Demand	Standard Op. Procedure	Preventive Maintenance	Fix as Required												TOTAL POINTS
			1	0	1	1	1	1	0											
1																				

WATER SERVICE RATING INSTRUCTIONS

1 point	if faucets and valves are in good repair.
0 points	if faucets and valves leak externally or internally.
1 point	if definite standard operating procedures are used. These should be written and posted.
1 point	if a definite preventive maintenance schedule is followed.
0 points	if equipment is maintained or repaired only when it breaks down.
1 point	if there is no equipment that uses once-through cooling water and discharges to sewer.
1 point	if water-consuming equipment is turned off when not in use.

SELF-EVALUATING CHECKLIST FOR WATER

EVALUATOR _____

DATE _____

UNIT NAME _____

SHEET NO. _____

WATER CONDITIONS

NO.	LOCATION	RATING VALUE MAX. = 5	No Faucet Leaks	Faucet Leaks	Standard Op. Procedure	Preventive Maintenance	Fix as Required	No Equip. Use Water Once	Equipment Off											TOTAL POINTS
			1	0	1	1	0	1	1											
1																				

PROCESS HEATING RATING INSTRUCTIONS

1 point   if the flue gas waste heat from processing equipment is extracted to heat relatively low temperature makeup, process and space heating water.

2 points   if all high-temperature piping, ovens, dryers, tanks and processing equipment are covered with suitable insulating material. The insulation must not be wet, crumbly or cracked.

0 points   if insulation is in poor condition with sections missing, broken, wet, crumbly or cracked.

1 point   if definite standard operating procedures are used. These should be written and posted near the control panel.

1 point   if gas-heated equipment is checked for combustion efficiency on a regular basis.

1 point   if a definite preventive maintenance schedule is followed.

0 points   if equipment is maintained or repaired only when it breaks down.

SELF-EVALUATING CHECKLIST FOR PROCESS HEATING

EVALUATOR _____

DATE _____

UNIT NAME _____

SHEET NO. _____

NO.	LOCATION	RATING VALUE MAX. = 6	Flue Gas Waste Heat	High Temp. Areas Insulated	Insulation Poor	Exhaust Process Air	Standard Op. Procedure	Combustion Efficiency	Preventive Maintenance	Fix as Required											TOTAL POINTS
			1	2	0	1	1	1	1	0										
1																				

PROCESS HEATING CONDITIONS

VEHICLE OPERATIONS/MAINTENANCE RATING INSTRUCTIONS

2 points   if a driver training course in economical operation is utilized.

2 points   if the fueling of vehicles is supervised and controlled.

2 points   for the maintenance of vehicles on a scheduled basis.

3 points   if there is an operating program on van pool and/or car pool.

0 points   if vehicles are observed operating with tires underinflated.

1 point   if a vehicle operating schedule procedure is in effect.

0 points   if vehicles are dispatched without regard for optimum use.

1 point   if mileage and fuel consumption data is available to all drivers.

SELF-EVALUATING CHECKLIST FOR VEHICLE OPERATIONS/MAINTENANCE

EVALUATOR_____

DATE_____

UNIT
NAME _____

SHEET NO. _____

NO.	LOCATION	RATING VALUE MAX. = 10	Economical Training	Supervised Fueling	Scheduled Maintenance	Van/Car Pool	Underinflated Tires	Dispatch Schedules	No Display Schedules	Mileage/Fuel Data											TOTAL POINTS
		VEHICLE OPERATIONS/MAINTENANCE CONDITIONS																			
			2	2	2	3	0	1	0	1											
1																					

11

Case Study:
A Complicated
and Successful
Energy Retrofit Program

INTRODUCTION

The Lone Star Gas Corporate headquarters, in Dallas, Texas, consists of a five building, 355,000 square foot office complex with buildings ranging in age from 14-60 years old between 20,000-100,000 square feet (SF), and with multiple HVAC systems served by two central plants. The company was facing the inevitable increase in electric utility costs knowing that two reactors at a new construction nuclear power plant would be coming on line during the next five years. By taking a proactive stance, a six month detailed energy audit commenced after a thorough internal energy analysis was conducted. The result of the audit was a multi-year $1.7 million retrofit project encompassing nearly 20 major items would be implemented over three years. Total energy and cost avoidance savings were calculated to yield a simple pay back of 1.5 years and a cash pay out of 3.5 years. Total energy reductions of 32% were achieved and the predicted economics realized.

The various projects involved the application of nearly 18,000 square feet of window tinting, a total facility relamping and efficiency improvement project, installation of more efficient filtration

Presented at 17th World Energy Engineering Congress by Martin A. Kimball, CDSM, CEM, FMA

systems, installation of control valves on the chillwater system, installation of a building automation system, installation of a plate heat exchanger for hydronic free-cooling, isolation of after-hours and 24-hour cooling loads on a separate loop, isolation and conversion of 24-hour steam requirements to reduce excess boiler capacity and run time, improvement of return air systems, the replacement and increase of cooling tower capacity, implement a preventive maintenance program, and improved operating procedures that focused on demand side management without thermal storage. The combined results of these single projects enabled the facility to remove and not replace 25% of the physical plant cooling equipment (one single-effect steam absorber) upon reaching the end of its expected life.

Project costs and avoided savings were tracked monthly throughout the three year period. Additional energy and cost avoidance tracking for two more years was completed. In five years, a positive cash flow of more than $550,00 is only 9% less than the original projection. Improved employee comfort and enhanced space conditions have returned significant benefits to the work force of nearly one thousand people occupying this facility.

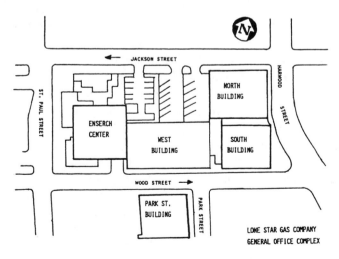

Figure 11-1. General Office Complex Layout

BACKGROUND

The general office complex (GOC) described in the abstract above consists of five buildings, as shown in Figure I above. The North, South, West, and Park St. buildings total approximately 255,000 SF and are served by one central plant located in the West building, until recently. The North and South buildings, 160,000 SF, pre-date central plant systems due to original construction periods from 1924-1933. A penthouse equipment room, using single effect absorption chillers, a 2-pipe hydronic exterior zone system with nearly 600 air induction units, and "state of the art" variable air volume (VAV) technology was installed in the North and South buildings in 1949.

During mid 1960's construction of the West Building, the South building HVAC loads were connected to the newer plant and part of the 1949 equipment was abandoned in place. The 80,000 SF West building consisted of a 3-pipe hydronic induction system for the exterior zone, and constant volume air handlers for the interior zone. This new plant was designed with cogeneration, using waste heat to operate 717 tons of single effect steam absorption.

Ten years later, the North building HVAC load plus the conversion-addition of the 20,000 SF two story Park St. building were added to West building plant when the cogeneration process was removed due to costly operations. The central plant capacity was nearly doubled by adding 650 tons of single effect steam absorption equipment, two additional 350 HP low pressure steam boilers to replace the waste heat source and power all four absorbers, and the primary chillwater loop modified with a "piggyback" main for the two additional secondary building loops.

The plant then consisted of four absorption chillers of 1367 tons on a primary chillwater loop with seven separate secondary loops serving the various zones in the Park St., West, North, and South buildings, and approximately 1700 tons of cooling tower capacity piped as two separate condenser water loops. (See Figure 11-2.) The "computer loop" was added as cooling requirements exceeded original design capacity for that floor. A separate 160 ton centrifugal chiller was later added to the computer loop to isolate and stabilize the load from the rest of the facility.

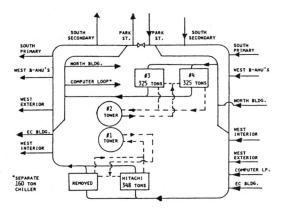

Fig. 11-2. Main Chill Water Diagram.

Only the four Park St. building air handlers (AHUs), of 33 primary AHUs, had any chillwater flow control devices installed. Therefore, nearly 90% of the cooling load consisted of "wild coils." Lastly, Figure 11-2 shows the replacement of one original chiller with a double-effect direct fired absorption chiller, and the complete removal of the other original 372-ton single-effect steam absorber.

The "EC bldg" loop was installed for contingency operations only, six years after the 1981 construction of the 100,000 SF addition to the headquarters office complex. The building was purposely designed to have it's own central plant with 300 tons of centrifugal chillers, hydronic heating, and a VAV system on each floor. Each AHU has a 3-way flow control valve on the chillwater coil. However, no spare cooling capacity was designed into the facility.

PROBLEM

The multiple modifications to the West building central plant presented numerous HVAC climate control problems due to the wild coils. Quite often, two opposing chillers serving the original and "piggyback" loop had to be operated to adequately supply all secondary loops with enough water flow to the suction side of the pumps, even though the actual cooling load was less than 40% of both machines. Additionally, all four absorption chillers were needed for peak summer cooling.

The plant was operated nearly 24 hours a day due to the inability to control secondary zone climate conditions. This practice was also carried over from earlier 24 hour cogeneration operations. Additionally, the house cleaning staff consisted entirely of company employees, and they were provided with conditioned spaces each night until 1:30 a.m.

Anticipation of higher utility costs, the continuing problems with unbalanced water flows, a significant amount of hands-on plant operation, and the near future necessity to replace the remaining original 22-year-old steam absorber created a high priority for determining a course of action for the physical plant mechanical systems.

DEVELOPING A PLAN: REALISTIC GOALS

All energy consumption and cost history was analyzed to determine a present benchmark from which future energy goals were to be achieved. The primary steam plant, serving four buildings, operated on 461,000 BTU/SF and averaged 7.29 watts/SF of electrical load for total utility operating cost of $3.26/SF for 1986. Meanwhile, the EC building operated on nearly 200,000 BTU/SF, averaged 3.0 watts/SF at a cost of $1.25/SF.

Higher costs and consumption for the larger plant were initially attributed to greater hours of operation to support the corporate data center cooling requirements, the use of single-effect steam absorption instead of electric centrifugals, and the inherent design of the primary and eight secondary chillwater loops each with a separate pump.

The decision was made to retain a professional energy consultant to assist the company in achieving it's goals. Three primary objectives of the goal were stated as:

1. Decide the tonnage size needed when Chiller #1 (372-ton absorber) is replaced.

2. Recommend a broad strategic five year plan for implementing energy and cost saving retrofits.

3. Provide recommendations for automating plant HVAC equipment over five years.

4. Provide facility occupants with a comfortable environment, and potentially cut energy consumption by 50%.

PHASE ONE: A DETAILED ENERGY AUDIT

A nationally experienced energy consultant in the Dallas area was retained. A detailed energy audit was conducted over a five month period in the first half of 1988, focusing on every aspect of the facility that required energy consumption or resulted in the need for space conditioning. Numbers of people, personal computer equipment, fax machines, copying machines, soft drink machines, coffee makers, light fixtures, document shredders, typewriters, etc., were counted and inventoried on a floor by floor basis. Detailed information on the building envelope regarding square footage of curtain wall, windows, roofing composition, orientation, insulation values, and shading coefficients were determined. All HVAC equipment, elevator machinery, miscellaneous machinery, and a detailed energy consumption and expense history for 1987 were provided to the consultant.

AUDIT RESULTS AND RECOMMENDATIONS

A computer model of the GOC operations was run on the TRANE Trace mainframe application. The output was compared to a different model that included the retrofit recommendations of the consultant. As expected, the potential to reduce annual consumption by more than 50% was predicted. The consultant's recommendations focused on more than 20 major items. The simple payback method indicated a 1.5 year return on investment (ROI) for a $1.2 million capital program.

The facilities staff further utilized the consultant's skills to develop a four-year retrofit schedule. Specific items addressed were grouped as appropriate, and a budget plan was established. Initial internal evaluations, using a discounted cash flow model (DCF), indicated that a $1.2 million investment would generate a ROI of more than 90% over a 15 year period.

The project schedule included:

1988:
1. Clean 56 AHU coils + 900 induction units. Upgrade filters.
2. Improve pneumatic control air dryer systems.
3. Isolate or convert non-HVAC steam loads from boilers.
4. Retrofit EC boiler with temperature reset controls.
5. Balance West physical plant ventilation air systems.
6. Replace incandescent w/compact fluorescent lights.
7. Replace outdoor quartz lights with high pressure sodium.

1989:
1. Install 18,000 SF of window tinting.
2. Air balance & upgrade EC building VAV system controls.
3. Refurbish 600+ induction units with new controls.
4. Relamp and upgrade facility office lighting.
5. Remove chiller #1 and modify expansion tank system.

1990:
1. Install Hydronic economizer and flow control valves on all AHU's.
2. Install new fancoil interior zone system to South Bldg.
3. Building Automation Systems (BAS) 75% complete.

1991
1. Complete South bldg. project.
2. Complete all building automation systems work.
3. Train plant staff on advanced BAS operations.
4. Implement automated work-order maintenance system.

The corporate training staff was used to further showcase the project's potential by developing a 23 minute video presentation for Senior management. Approval was granted to commence immediately, and increase the 1988 budget in less than 45 minutes.

1988 PROJECTS AND RESULTS

By mid-October, significant work was underway for installing 18,000 SF of window tinting and cleaning all AHU coils and upgrading coils. November 1988 GOC energy consumption was 16% <u>less</u> than November 1987. The increased coil surface efficiency created additional comfort control problems resulting from better heat transfer.

More than $20,000 in lighting retrofits of interior and exterior incandescent lighting was completed by the end of December. The majority of the interior lights are on 24 hours a day year round, in stairwells and elevator lobbies. The calculated ROI for this portion was less than 12 months.

Steam humidification control in several computer room Liebert air conditioning units were replaced with electric quartz elements. The facility domestic hot water (steam) heater was modified, by adding a separate boiler to the system. The steam heat was valved off as a backup source.

The additional air dryer for plant control air had immediate results. While the original unit was sufficient for system capacity, the boiler room environment routinely had high humidity from occasional deareator pop-offs. Once the control air system was dried out, various plant systems began operating more smoothly.

1989 PROJECTS AND RESULTS

The EC building (12 stories) required a complete air balance as well as upgrades to the building control system. Many VAV units were manually locked in various positions. This "solution" was due in part to earlier commissioning of the structure. The ceiling plenum return air path to the AHU room required 100% of the return air to be drawn between two areas with little clearance beneath I-beams. A resulting affect was that hallway ceiling tiles located directed beneath the AHU room intake were lifted from the grid as the air flow took the path of least resistance. A quick fix-it solution was to install several 2' × 2' egg crate open tiles. This solved one problem yet short circuited the design of the system causing the hallway to be used as a return air plenum.

To remedy the above situation, new tiles were installed using clamps to prevent lifting. A complete air balance, better placement of the static pressure sensor in the main air ducts, and DDC automation of the 3-way chill water valves improved comfort problems. Cooling tower, chiller, pump controls, kW demand metering, and integration for optimizing building start/stop times was begun.

More than 5000 office fluorescent lighting fixtures received "special" attention. A very laborious process, a detailed lighting survey, with occupant interviews, led to the development of a complicated plan. Every fixture was relamped, cleaned, and had new acrylic lenses installed. (The fixtures had not been cleaned for over 15 years!) Due to a high fixture density, about 1100 were removed. This later involved additional work to properly dispose of PCB laden ballasts. Many of the remaining fixtures were either delamped to two lamps per fixture and/or relocated. The resulting ROI for the entire lighting project, including the purchase of 1100 ceiling tiles, was under two years. Nearly 300 kW, or 32%, of the beginning lighting energy was removed due to increased efficiencies. Additionally, average light levels throughout the GOC improved by 25%.

Detailed plans to replace temperature control devices on the 600 (35-year-old) induction units, and repipe nearly 300 3-pipe induction units into somewhat of a 4-pipe zoned system was canceled. Contractor bids were more than double the original estimates due to excessive risk in working on active systems requiring repeated drain down and fill times, unknown variables about the integrity of the aged chillwater piping, and having to work in occupied building and protect furniture and equipment from potential water damage

By the beginning of August 1989, the summer peak cooling load served by the West bldg. plant was significantly reduced. A combination of the cleaner coils, better heat transfer, reduced lighting energy load, window tinting, and improved plant controls eliminated the need for the fourth absorption chiller (as a back up). Hence, this 372 -ton chiller was removed and no replacement was necessary.

By the end of 1989, annual energy consumption had been reduced by more than 25% from 1987. Total project costs to date were $401,000. Avoided costs of more than $417,000 gave the project a positive cash flow of $16,000 even though none of the projects requiring major construction and piping changes had yet begun.

Excellent savings and improved energy efficiencies had now been accomplished in two years with a little over one-third of the construction budget. However, more significant changes and detailed planning were necessary as the rest of the projects became the main thrust of the strategic plan for the central plant and future operations.

1990 PROJECTS AND RESULTS

Two significant projects requiring nearly $525,000 took place in late 1990 and early 1991. These projects included 1) installing 2-way flow control valves on the major AHU's and connecting 5 of 9 secondary loops to a new plate and frame heat exchanger for free cooling, and 2) extending the "computer loop," supplied by chiller #5, throughout the North, South, and West buildings. This extension would enable future 24 hour, high density HVAC loads to receive cooling separate from the primary systems. Secondly, 24 hour penthouse cooling would be needed for future electronic elevator control equipment as modernization occurs and for present elevator equipment in the West building penthouse when HVAC services were to be cut back on nights and weekends.

Figure 11-3 simplifies the piping schematic involving the free cooling plate heat exchanger installation. Only those loops or building zones requiring cooling year round were added. The computer loop was added for two reasons. First, the opportunity arose to maximize cooling potential in the winter months without the use of chiller #5 when outdoor temperatures drop below 45 degrees F. Second, in the event of extremely cold weather, the heat rejection from the computer loop may assist in preventing freeze damage to the cooling tower cells in use. The frame size for the heat exchanger was specified so that future addition of the EC ("ENSERCH Center" in Figure 11-3) building winter cooling load could be added. This would finally enable the shut down of all chillers and absorbers between 1000-1500 hours a year.

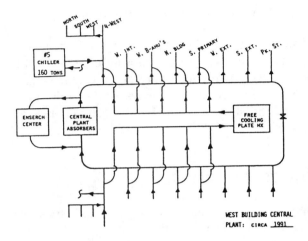

Fig. 11-3. Free Cooling Heat Exchanger

The staging of pipe sections, valves, rigging equipment, and flanges was crucial for this addition, because it was also coupled with the installation of 2-way flow control valves on the major AHU's. Installation requirements mandated that the computer loop remain in operation and be isolated from the rest of the chill water system that was drained. One condenser water loop also remained in operation.

The 4-day Thanksgiving weekend provided enough time to complete the drain down, install necessary valves, replace original non-functional valves, and install cross over valves on the condenser water loop. Cross over valving was added to increase system flexibility during cooling tower replacement in 1991. Figure 11-4 shows this arrangement which allows the plate heat exchanger to operate on either condenser water loop while absorption machines are operated on the other.

By the end of 1990, an additional $450,000 in project expenses were incurred covering year end work. Total program costs of $856,000 minus avoided utility expenses netted a negative cash flow of $110,000. Even though the cash flow model change of $126,000 from a positive to a negative number occurred, the amount was still

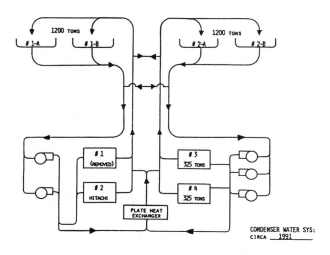

Fig. 11-4. Condenser Water Loops.

$100,000 <u>less</u> than was originally estimated and acceptable to the company.

1991 PROJECTS AND RESULTS

The project involving the addition of an interior zone system to the South Building was canceled, again due to extremely prohibitive construction costs. Instead, the focus remained on completing piping connections for the plate heat exchanger loop and gaining automated control of all the new 2-way valves and heat exchanger valving.

The original wood cross-flow cooling towers were replaced in 1991. This project was separate from the original proposal submitted by the consultant, yet dove tailed nicely into the overall strategic plan for the facility.

Continuous leakage around the tower basins, some structural steel decay on the submerged I-beam foundations in the basin and the fan gearbox supports resulted in part due to extensive operations this equipment received over the years. After reviewing the cost estimates required to make a 25-year-old tower operate like a brand new 25-year-old tower (i.e., single speed motors and wooden basins), a complete tower replacement decision received management approval.

The 1700 tons of original tower capacity was replaced and increased to 2400 tons (four 600 ton cells) of fiberglass counter flow towers with 2-speed fan motors. On a side note, nearly 81,000 lbs. of operating weight for the wooden towers became less than 25,000 lbs with the new towers.

The increase in sizing provided future up-sizing from 325-ton single effect steam absorption equipment to 400- to 450-ton replacement double effect direct-fired absorption equipment. This up-sizing would be crucial since the primary chillwater machinery now consisted of only three instead of four machines.

Analysis of total water consumption showed that nearly 40% of annual usage was associated with the cooling towers. Local rates provided excellent cost savings if more than 10% of total consumption is "9" type of use. The tower make up and bleed off water was separately metered to get credit for this discount. Total costs of less than $3,500 resulted in a simple ROI of less than four months.

At year's end, total project costs were nearly $1.1 million. Nearly $1.2 million in avoided energy costs and 30% energy reductions were achieved in only 36 months.

1992 PROJECTS AND RESULTS

Detailed energy consumption history was used for operations analysis, utility cost forecasting, and project reporting. Changes to the operation of both central plants also reduced operating costs. Primarily, the 300 tons of electric chillers in the EC building are shut down during business hours. Chillwater is provided from the West building absorption plant. While energy consumption has risen, operating costs have been reduced by eliminating the daytime energy and electrical demand costs associated with electric equipment. These machines are only run after hours when the building kW loads had been significantly lowered from the shut down of AHU's and office lighting.

The facilities HVAC operations, excluding minimal 24 hour loads, have been routinely shut down at night and on weekends. The plant staff is now accustomed to predicting operating patterns based on weather conditions. Space conditioning complaints continued to decrease and occupant morale is perceived as continually improving.

The construction budget for training, completing the BAS system, and automating maintenance were already included in previous years costs.

FIVE-YEAR PROJECT SUMMARY

Cooling tower replacement costs of nearly $625,000 were included with the $1.1 million multi-year retrofit projects in order to track additional avoided costs and determine a combined cash flow model. While slightly more than $1.7 million had been spent in 3.75 years, a net positive cash flow of $70,000 was generated by the end of 1992. Annual utility costs had dropped by 15% since 1987 and total energy consumption decreased by 32%. Figure 11-5 presents these trends graphically.

PHASE TWO: THE NEXT STEP

The ability to incorporate strategic design elements into the physical plant modifications has paid handsome dividends. As of mid-summer 1994, additional HVAC plant modifications have included:

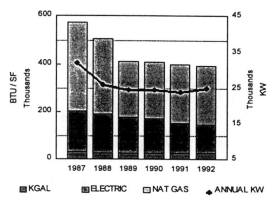

Btu/SF and kW History

Fig. 11-5. Energy Consumption History.

1. Complete DDC automation and improved comfort control of the two story Park Street building.

2. Replacement of two 325-ton single effect steam absorption units with one 400-ton and one 450-ton double-effect direct fired TRANE chillers.

3. Adding the EC building winter cooling load to the free cooling plate heat exchanger.

4. Replacement of several 40$^+$-year-old AHU's, allowing better space control and full DDC automation in the South building.

Objectives for the immediate five year future include the following:

1. Remove all three 350 HP low pressure steam boilers and replace with a series of smaller, more efficient units arranged in a step fired sequence.

2. Removal of 300 tons of electric centrifugals in the EC building and replace with gas equipment.

3. Add 24-hour air conditioning to the two unconditioned elevator equipment rooms during modernization.

SUMMARY

Many facilities possess the potential for tremendous energy conservation opportunities (ECO's). While the dedication and knowledge of the in-house staff can accomplish a lot of individual ECO's, the formulation of an extensive multi-year master plan requires the skill of a knowledgeable professional. Businesses and owners must be willing to embrace the potential when a "reach for the sky" can take you all the way to the stars.

Lone Star Gas did not proceed blindly into this program. Strong supporting evidence regarding the performance history of the TRANE Trace program and the professional energy consultant's experience was meticulously scrutinized. Upon our complete satisfaction, project approval was granted.

Energy reductions of more than 50% were not achieved because all of the original recommendations were not completed. Should the company have agreed to proceed with the seemingly cost prohibitive construction items, then an additional 18% energy reduction may have occurred. In turn, the project ROI and cash pay out may not have been affected at all. However, the 32% reduction and repeated cost reduction in annual energy costs is a clear example that energy engineering can return significant dividends to the user.

12

Case Study: Comprehensive Facility Energy Assessment Using FEDS

INTRODUCTION

The energy savings and demand reduction opportunities at the Army's National Training Center at Fort Irwin, California, were evaluated. The Fort Irwin analysis made use of the recently developed Facility Energy Decision Screening (FEDS) System Level-2 software tool. FEDS is a systematic, technology-neutral, and fuel-neutral approach to evaluating energy savings opportunities at large facilities. FEDS analyzes most major building end uses (e.g., heating, cooling, lighting, ventilation, and service hot water), including interactive effects (e.g., the effect of a lighting technology on heating and cooling loads). FEDS output provides specific cost, energy (and demand) charges, and life-cycle cost (LCC) information, by cost-effective energy resource opportunities (EROs). The remaining end uses common to large facilities (e.g., motors, transmission and distribution, vehicles) are analyzed using manual calculation methods.

The present value (PV) of the installed cost of all EROs constituting the minimum LCC efficiency resource (i.e., cost-effective) at Fort Irwin is approximately $23.9 million in 1994 dollars (1994$).

Presented at 17th World Energy Engineering Congress by Donald L. Hadley, John M. Keller, Eric E. Richman and Rene Quinones.

The PV of the energy and demand, operations and maintenance (O&M), and replacement savings associated with this investment is approximately $87.3 million, for an overall NPV of $63.6 million.

This chapter will describe the FEDS process and present detailed results of the comprehensive energy resource assessment conducted at Fort Irwin.

WHAT IS FEDS?

The number of conceivable energy conservation measures, fuel-switching opportunities, and renewable energy projects at a federal site is very large. The Pacific Northwest Laboratory (PNL) uses two methods to select, evaluate, and prioritize these energy resource opportunities (EROs). The first is the Facility Energy Decision Screening (FEDS) Model used for most building end uses. The second is a manual process to evaluate all remaining end uses.

FEDS is a multilevel software tool designed to provide a comprehensive approach to fuel-neutral, technology-independent integrated (energy) resource planning and acquisition. FEDS currently has two levels—Level-1 and Level-2. Level-1 is a menu-driven DOS-based software program designed for facility energy managers as a screening tool. Level-1 assesses the likelihood of cost-effective energy projects based on high-level facility inputs and numerous assumptions. The output of Level-1 is used to assess a facility's overall energy conservation potential from the perspectives of potential energy savings, potential cost savings, and estimated investment requirement.

Level-2 is also a DOS-based software program that can be used by facility energy managers to identify, characterize, and assess individual energy projects. However, Level-2 goes to the next level of detail, providing specific information on energy and cost savings, as well as the estimated investment requirement for specific technology retrofits. Level-2 is the appropriate analysis to follow positive Level-1 results; typically, a Level-2 input file can be initiated from a Level-1 input file. Level-2 allows the user to enter facility-specific data inputs to replace the inferred default values from Level-1. These inputs form "building sets," which are groups of buildings similar in use, age, construction type, fuel use, fuel availability, or other definable characteristics. By developing building sets based on detailed facility data, Level-2 tailors the analysis to the facility and provides more accurate and detailed economic findings.

At this point in the software development, Level-1 and Level-2 analyze most major building end uses (heating, cooling, lighting, ventilation, and service hot water) including their interactive effects (e.g., the effect of a lighting technology on heating and cooling loads), providing specific cost, energy (and demand changes), and LCC information, by cost-effective technology.

The second method PNL addresses those end uses not analyzed by the FEDS software. This analytical approach is a three-step manual-calculation (hereafter referred to as "Manual") process which has been developed by PNL to make energy resource opportunity (ERO) selection, evaluation, and prioritization manageable. The steps are:

- Preliminary Screening. Select promising EROs from a master list, considering the site's mission, building stock, end-use equipment characteristics, utility characteristics, climate, energy costs, and other local conditions that affect ERO viability, and recommendations from site staff.

- Cost and Performance Analysis. Establish, with a reasonable degree of accuracy, the technical and economic feasibility of each ERO that passed the preliminary screening. Perform an analysis comparing the operating and economic performance of the existing equipment and the ERO. Where applicable, include impacts on energy security and the environment in the analysis.

- Life-Cycle Cost Analysis and Prioritization. Perform an LCC analysis and rank EROs by net present value (NPV), so that a package with the optimal return on investment can be defined. If any utility cost-sharing or rebate programs exist, they can be included within this evaluation step.

All federal agencies are required to evaluate the LCC of alternative technologies when making energy investments. The LCC analysis and prioritization step used in both the Level-2 and manual methods is required by, and complies with, federal law [1]. An LCC evaluation computes the total long-run costs of alternative actions and identifies the action that maximizes the NPV of the energy investment.

FORT IRWIN CHARACTERISTICS

Fort Irwin is a roughly 1000-mi^2 U.S. Army Forces Command (FORSCOM) facility situated in the Mojave Desert approximately 37 miles northeast of Barstow, California, and south of Death Valley. The main cantonment area is located near the southeastern portion of Fort Irwin. The Fort's primary mission is to operate the National Training Center (NTC). The NTC is a support facility for training of troops normally stationed at other posts throughout the United States. A total of twelve 28-day training rotations are scheduled each year. The Fort mission results in erratic energy consumption because a large portion of the Fort population is transient, moving on- and off-site as dictated by the training schedules.

The climate at Fort Irwin is classified as "high desert," with an average annual rainfall of 2.5 in., most of which falls between December and February. Summer maximum temperatures are around 104°F, and winter minimum temperatures are around 29°F. Annual heating and cooling degree-days (base 65°F) are 2,547 and 2,272, respectively.

Building Characterization

Roughly 842 commercial buildings (not including schools) with a floor area of 3,439,606 ft^2 are reported in the Fort Irwin Real Property Data Base (RPL). An additional 732 housing buildings (1636 units, not including General's Quarters) with a reported area of 2,961,830 ft^2 contribute to the Fort's total building area of 6,401,436 ft^2.

Based on the RPL, the facilities at Fort Irwin may be divided into 36 building types. These building types are created by combining facilities of different facility description codes (as provided in the RPL) into larger categories with similar energy usage. This procedure minimizes the number of building types while preserving any unique or unusual building characteristics that have an effect on energy consumption.

Family housing (2.9 million ft^2) is the single largest category by square footage at Fort Irwin, followed by barracks, administration, motor pools, warehouses, manufacture administration, and general shops. These building types account for more than 80% of the total building stock at Fort Irwin.

Commercial buildings are a mix of older wood frame construction and newer stone/brick construction, with some metal frame and curtain wall construction. Family housing is primarily wood frame construction with varying levels of insulation in the walls or ceilings.

Electric Utility Service Characterization

Electric service to Fort Irwin is provided by Southern California Edison (SCE). Distribution on the site consists of five 12-kV transmission lines from two substations. Both the transmission and distribution systems are overhead line systems for most of the commercial areas. Most family housing areas are supplied by underground lines.

The Fort Irwin electric system has approximately 610 transformers, with a total estimated nameplate capacity of more than 35,000 kVA. The losses associated with transformer operation are estimated at an average level of 272 kW, for a total yearly loss of 2,382 MWh.

Water and Sewer Service Characterization

Water is provided to Fort Irwin by eleven operating wells and six booster pumps. Eight of the wells are located in and around the main commercial area. The remaining three wells are to the northeast toward the Bicycle Lake Basin. The aggregate capacity of the well pumps is reported to be 4,595 gpm with an aggregated load of 1,040 hp. Booster pumps are also in operation with each well system at a total effective load of approximately 1,050 hp.

Fort Irwin also has a demineralized water production system. The demineralized water is supplied to all occupied facilities through a separate distribution system. Demineralization is accomplished at the Reverse Osmosis (RO) Plant. The current major aggregate load of the equipment at the RO Plant is approximately 255 hp.

An on-site system provides sewage treatment for Fort Irwin, including the central Fort facilities and family housing. The current major aggregate load of the equipment at the sewage treatment plant is approximately 210 hp.

Street and Parking Lot lighting Characterization

Exterior street and parking lot lighting at Fort Irwin comprises a variety of lighting types and capacities including high-pressure sodium (HPS), mercury vapor (mercury), and metal halide (MH). Accurate records of quantities and capacities were not available from Fort Irwin sources. A manual lighting count was completed that included all street and parking lot lights in and around the commercial, family housing, and nearby field station areas. Also included in the count were the perimeter lights at the ordnance storage area. This count, along with some lamp procurement information collected at the Fort, was used to estimate street and lot lighting energy consumption.

The total streetlighting demand was calculated to be 338 kW. Assuming a conservative 10 hours of operation per day (many will operate longer but some are used only during specific activity), this demand is equivalent to a total of 1,232 MWh/yr.

ENERGY SOURCE CHARACTERISTICS

Energy sources used by Fort Irwin include electricity, propane, and vehicle fuel. Fuel oil is used in small quantities primarily for backup generator systems and is not considered in this assessment. Table 12-1 shows a summation of the sample yearly energy consumption and cost at Fort Irwin for all facilities including family housing. For each energy type, the yearly total is shown in units appropriate to the energy as well as a common unit as a basis of comparison. The total consumption values are based on the usage values chosen for analysis in this report as typical current yearly usage. This sample year is based on the best available data gathered from 1991 through April 1993 consumption levels. These aggregations of various billing consumption amounts are considered representative of normal Fort Irwin operational energy consumption.

TABLE 12-1. Annual Energy Consumption and Energy Cost

Annual Energy Type	Annual Consumption Consumption	Percentage of (MBtu)	Annual Cost Total Energy	Percent of (1994$)	Total Cost
Electricity	72,860 MWh	248,645	11.9	6,271,513	52.8
Propane	2,369,487 gal	225,101	10.8	1,120,293	9.5
Gasoline	446,098 gal	55,762	2.7	370,261	3.1
Diesel	3,718,042 gal	515,692	24.7	2,491,088	21.0
JP-4	770,500 gal	184,646	8.9	577,875	4.9
JP-8	1,367,750 gal	853,954	41.0	1,025,813	8.7
Totals	—	2,083,800	100.0	11,856,843	100.0

Electric Supply Source Description

Electricity is supplied to Fort Irwin by Southern California Edison (SCE) and delivered to the Tiefort Terminal Station via a 115-kV transmission line. A subtransmission line rated at 34.5 kV connects this station to the Fort Irwin substations. Fort Irwin is supplied under rate schedule TOU-8 with an additional incremental sales rate agreement covering additional electricity usage above pre-set limits.

Propane Supply Source Description

Propane is supplied to Fort Irwin through a competitive contract and delivered to a central propane storage area consisting of four 30,000-gal tanks. The propane is distributed through 8-, 6-, 4-, and 2-in. lines throughout the Fort and is used primarily for heating, cooking, and water heating. The current propane rate is $0.473/gal.

Vehicle Fuel Supply Source Description

Fuels for vehicle use are supplied to Fort Irwin by various contract suppliers and stored at point-of-use locations in commercial areas. The primary fuels are gasoline, diesel, JP-4 (aviation fuel), and JP-8 (combination aviation/ground fuel). Current fuel prices, in dollars per gallon, are gasoline at $0.83, diesel at $0.67, JP-4 at $0.75, and JP-8 at $0.75.

FEDS-LEVEL 2 SUMMARY RESULTS

The present value (PV) of the installed cost of all EROs constituting the minimum 1 CC efficiency resource (i.e., cost-effective) at Fort Irwin is approximately $23.9 million in 1994 dollars (1994$). The PV of the energy and demand, O&M, and replacement savings associated with this investment is approximately $87.3 million, for an overall NPV of $63.6 million.

Table 12-2 provides a breakdown and summary of the cost-effective EROs at Fort Irwin. The O&M savings are a reflection of the incremental cost difference between the cost of maintaining the existing equipment and that of maintaining new or retrofitted equipment. Because maintenance costs of new or retrofitted equipment are often the same as the costs to maintain the existing equipment, this incremental maintenance cost is often zero.

TABLE 12-2. Summary of the Cost-Effective EROs (1994$)[a]

ERO Category	Present Value of Installed Cost	Present Value of Energy and Demand Savings	Present Value of O&M Savings	Present Value of Replacement Savings	Present Value of Total Savings	Total Net Present Value
Lights (Level–2)	4,393,028	17,464,385	0	4,184,358	21,648,743	17,255,714
Vehicles	2,047,000	5,662,859	6,475,790	0	12,138,649	10,091,649
Envelope	1,400,907	11,619,936	0	−789,727	10,830,209	9,429,302
Roof (Level–2)	2,005,349	8,131,276	0	0	8,131,276	6,125,922
Fam. Hsg. HVAC	7,086,917	12,291,871	281,903	−241,994	12,331,780	5,244,863
Lighting Controls	180,827	2,512,676	719,268	0	3,231,943	3,051,116
Motors	1,362,331	4,051,014	4,133	−504,490	3,542,390	2,180,059
HVAC	279,627	2,565,025	0	−126,243	2,438,782	2,159,155
Trans. & Dist.	2,543,519	2,242,172	−109	2,147,346	4,389,410	1,845,890
Hot Water (Level–2)	188,447	1,743,372	0	0	1,743,372	1,554924
Wall (Level–2)	907,261	1,840,887	0	0	1,840,887	933,622
Central Chillers	354,000	1,273,017	−25,831	0	1,247,186	893,186
DHW & A/C	118,124	1,001,673	−32,719	−53,507	915,446	797,322
Wells	210,500	718,156	4,305	51,131	764,981	554,481
A/C 90	539,429	−1,550	0	537,879	537,789	
Heating	235,202	700,930	0	22,215	723,146	487,944
Controls	150,400	532,652	0	−68,127	464,525	314,125
Cooling (Level–2)	165,900	274,395	0	0	274,395	108,496
Heating (Level–2)	13,016	45,366	0	0	45,366	32,352
Totals[b]	23,908,628	72,211,089	7,408,314	4,620,963	87,2474,447	63,625,347

Notes: (a) Data of this level of detail is not normally available from FEDS Level-2. All values from the Level-2 software are approximate, and are shown only to represent the magnitude of the savings from each end use. (b) These totals are the sum of the manual EROs and the output from the Level-2 software. They will not necessarily be the sum of the numbers above.

Table 12-3 presents a breakdown and summary of both the energy and demand savings for the first year and full implementation of the cost-effective energy resource at Fort Irwin.

For EROs analyzed by FEDS Level-2, lighting EROs represent the greatest efficiency resource, accounting for more than $17.3 million of the total $63.8 million NPV and $4.4 million of the total $24.7 million installed cost. The remaining ERO categories have NPVs ranging from $6.1 million to $0.9 million, except for cooling and heating EROs, which are only marginally cost-effective with NPVs of $108,500 and $32,400, respectively.

For non-building EROs, vehicles represent the greatest efficiency resource, accounting for $10.1 million of the total $63.8 million NPV and more than $2 million of the total $24.7 million installed cost. The remaining nonbuilding ERO categories have NPVs ranging from $9.4 million to $314,000.

For building EROs (analyzed by Level-2), the estimated annual electricity consumption at Fort Irwin is 89,100 MWh. Estimated electric demand is 30,100 kW. Full implementation of all electric EROs results in a reduction of 14,500 MWh and 3,600 kW. This represents a reduction of approximately 16% over total electricity consumption and 12% over site-wide demand. The estimated annual propane consumption at Fort Irwin is 209,100 MBtu. Full implementation of all propane EROs results in net conservation of 71,000 MBtu, which represents a net conservation of 34% of total consumption. The end uses of chilled water and district hot water were not broken out by fuel. The estimated annual chilled water use is 2 million ton-hours. Full implementation of all chilled water EROs results in a reduction of 331,000 ton-hours, or 16% of total consumption. The estimated annual district hot water use 9,200 MBtu. Full implementation of all district hot water EROs results in a reduction of 7,700 MBtu, or 83% of total consumption.-

For non-building EROs, the estimated annual electricity consumption at Fort Irwin is 79,800 MWh. Estimated electric demand is 399,000 kW-month (sum of the peak demands for each month). Full implementation of all electric EROs results in a reduction of 12,200 MWh and 58,000 kW-month, representing a reduction of approximately 15% over total electricity consumption and 14% over site-

TABLE 12-3. Summary of Energy and Demand Savings from EROs

ERO Category	First-Year Energy Savings (MBtu)	First-Year Demand Savings (kW-mo)	Implement Energy Savings (MBtu)	Full Implement Demand Savings (kW-mo)	Annualized Energy and Demand Savings (1994 $)
Lights (Level-2)	NA	NA	34,815	2,487	1,014,144
Fam. Hsg. HVAC	76,678	15,226	76,678	15,226	713,785
Envelope	21,862	17,099	21,862	17,099	674,766
Roof (Level-2)	NA	NA	45,939	621	472,181
Vehicles	14,638	-180	14,638	-180	328,840
Hot Water (Level-2)	NA	NA	40,609	20	242,457
Motors	7,814	7,343	7,814	7,343	235,241
HVAC	15,058	1,690	15,058	1,690	148,950
Lighting Controls	5,992	0	5,992	0	145,910
Trans. & Dist.	2,203	3,708	6,076	7,223	130,202
Wall (Level-2)	NA	NA	10,653	123	106,898
Central Chillers	1,099	2,110	1,099	2,110	73,924
DHW & A/C	4,345	2,198	4,345	2,198	58,167
Wells	0	1,097	0	1,097	41,703
Heating	4,682	727	4,713	742	40,703
Cooling (Level-2)	NA	NA	962	143	35,400
A/C	1,129	935	1,129	31,324	30,931
Controls	1,508	2,186	1,508	2,186	3,976
Heating (Level-2)	NA	NA	-71	0	
Totals:	156,813	54,331	293,623	61,256	4,529,501

wide demand. The estimated annual fossil fuel consumption (natural gas, No. 2 fuel oil, propane, gasoline, and diesel) at Fort Irwin is 823,800 MBtu. This total excludes any diesel and gasoline used for vehicles not addressed through EROs. Full implementation of all fossil fuel EROs results in conservation of 187,000 MBtu and a new load of 68,800 MBtu, for a net reduction of 118,600 MBtu. This represents conservation of 23% of total consumption, a new load of 8%, for an overall decrease of 14% in fossil fuel use.

ENERGY PROJECT IMPLEMENTATION AT FORT IRWIN

To meet its target of reducing overall energy consumption by 30% by the year 2005 (1985 baseline), Fort Irwin has developed a 5-year plan and is pursuing base-wide energy conservation. Sources of funding for implementing these energy conservation projects include the Department of Army's Energy and Conservation.

Investment Program (MILCON/ECIP) and the Federal Energy Management Program (FEMP). Another potential source of funds is the utility-sponsored demand-side management (DSM) programs. The FEDS Level-2 results are used to prioritize the most cost-effective energy projects by evaluating the projects' life-cycle costs, investment requirements, and the energy savings opportunities.

Five-Year Energy Plan

Fort Irwin has developed an extended five-year energy plan that provides a timeline for implementation of energy conservation projects and identifies potential funding mechanisms. Individual energy projects identified by the FEDS process have been folded into this plan. The five-year plan is extremely dynamic, responding to the annual cycle of available funds or changes in utility DSM programs.

The detailed spreadsheet format of the FEDS Level-2 output allows for a relatively easy identification of individual energy projects that can be implemented as time-phased projects targeted to available funding sources. Specific projects can be identified by disaggregating the results either by building or end-use category or by a specific retrofit technology applied across multiple end uses.

As part of this process, the FEDS results were used to identify five energy projects that were submitted for FY95 FEMP funding. These projects, shown in Table 12-4, met the program requirements for simple payback and savings-to-investment ratio.

Residential HVAC Evaluation

The FEDS process was also used to evaluate alternative scenarios for heating and cooling of the existing family housing. There are 1,637 family housing units, all with basically the same propane furnace and central air conditioning systems. The options evaluated included air-source heat pump, ground-source heat pump, LPG furnace and central air, natural gas furnace and central air, and gas-fired heat pump (still under development but included for comparison purposes).

Although natural gas is not currently available at Fort Irwin, HVAC options are included to compare the operating cost of natural gas and propane. The natural gas rate used in the analysis is an estimate based on information provided by Fort personnel and representatives of possible natural gas providers.

This ERO was analyzed manually because the Level-2 software cannot yet fully analyze EROs involving heat pumps (either air- or ground-source) or fuel switching from LPG to natural gas when natural gas is not available to the building. Therefore, all residential HVAC options were analyzed manually, using only the savings from the individual pieces of equipment.

The technical assumptions are as follows:

- The existing LPG furnaces have an average size of 50 KBtu/h (input) and efficiency of 70.5% AFUE. The existing air conditioners have an average size of 2.5 tons and efficiency of 8.0 SEER.

- The replacement equipment efficiencies are shown in Table 12-5.

- Existing energy consumption was calculated using previously developed energy use intensities (EUIs) [2]: 2.91 kWh/ft^2-yr for

TABLE 12-4. Energy Projects Identified for FEMP Funding

Project/Description	Initial Cost	Savings-to-Investment Ratio	Discounted Payback
Cross-connect chillers: Provide the capability to cross connect chiller equipment between two central energy plants. During periods of low loads, this would allow operation with only one chiller, improving overall plant efficiency.	$354,000	3.5	4.9 years
Reset chilled/condenser water temperature: Reset the chilled water and condenser water temperatures 2-4- higher/lower. Large water-cooled chillers are typically set for 45°F chilled water temperature and 95°F condenser water temperature. Most systems can operate at higher/lower temperatures without impacting cooling performance.	$1,000	5975	0.0 years
Install A/C desuperheaters: Install desuperheaters on air-cooled air conditioners in selected commercial buildings. This would improve overall A/C efficiency by about 15% and would supply the majority of the hot water needs during the cooling season.	$118,000	7.8	2.2 years
Replace existing residential domestic water heaters: Replace existing domestic hot water heaters, wrap tanks and piping with insulation, and lower tank temperature.	$270,000	6.4	1.6 years
Replace space heaters with LPG infrared heaters: Replace conventional space heaters with LPG infrared heaters in selected maintenance shops and motor pool buildings.	$30,000	6.3	2.8 years

TABLE 12-5.   Efficiency Ratings of Replacement HVAC Equipment

Replacement Equipment	Cooling Eff.	Heating Eff.
Minimum Compliance Air-Source Heat Pump	10.0 SEER	7.0 HSPF
High-Efficiency Air-Source Heat Pump	15.4 SEER	8.3 HSPF
Average-Efficiency Ground-Source Heat Pump	13.3 EER	2.8 COP
High-Efficiency Ground-Source Heat Pump	16.0 EER	3.5 COP
Minimum Compliance Furnace and A/C	10.0 SEER	78.0% AFUE
High-Efficiency Furnace and A/C	15.7 SEER	92.6% AFUE
Gas-Fired Heat Pump	1.1 COP	1.3 COP

Notes:

(1) Efficiencies for the LPG and natural gas furnace and central air conditioner options are assumed the same; the only difference for these two options is the price of fuel.

(2) Additional notes on ground-source heat pumps: 1) There are no efficiency standards for ground-source heat pumps, so an average efficiency unit was chosen to represent the minimum compliance case. 2) Because the ground temperature remains fairly constant, the given efficiencies are assumed to represent seasonal values (EER = SEER).

cooling and 26.37 kBtu/ft^2-yr for heating. For an average house size of 1,800 ft^2, the energy consumption is 5,238 kWh for cooling and 47.5 MBtu for heating per unit.

• Retrofit energy consumption is based on the actual equipment size and estimated run hours of each replacement unit to meet the same load as the existing equipment. The replacement equipment sizes are different from the existing equipment size in almost all cases because actual equipment was chosen for the retrofit options. Equipment sizes are given in Table 12-6.

• Operating hours for the existing equipment are based on the EUIs and equipment capacities as described above. Operating hours for the retrofit equipment are calculated from the existing equipment hours modified by the replacement equipment efficiencies and capacities.

The cost assumptions are as follows:

• The replacement equipment installed costs are shown in Table 12-7.

TABLE 12-6. Replacement Equipment Sizes

Replacement Equipment	Cooling Cap. (KBtu/h)	Heating Cap. (KBtu/h)
Minimum Compliance Air-Source Heat Pump	28.2	27.4
High-Efficiency Air-Source Heat Pump	29.0	29.0
Average-Efficiency Ground-Source Heat Pump	30.2	20.8
High-Efficiency Ground-Source Heat Pump	31.2	21.2
Minimum Compliance Furnace and A/C	28.6	40.0
High-Efficiency Furnace and A/C	30.8	37.0
Gas-Fired Heat Pump	36.0	53.5

- O&M costs are $75/yr for all air- and ground-source heat pump options, $85/yr for all furnace and air conditioner options (including the existing), and $105/yr for the gas-fired heat pump option.

- The cost of natural gas is assumed to be $3.50/MBtu.

TABLE 12-7. Installed Cost of Replacement Equipment

Replacement Equipment	Material (1994 $)	Labor (1994 $)
Minimum Compliance Air-Source Heat Pump	$2,180	$559
High-Efficiency Air-Source Heat Pump	$5,175	$559
Average-Efficiency Ground-Source Heat Pump	$3,000	$559
High-Efficiency Ground-Source Heat Pump	$3,770	$559
Minimum Compliance Furnace and A/C	$1,483	$468
High-Efficiency Furnace and A/C	$4,725	$468
Gas-Fired Heat Pump	$5,000	$750

Note: Material costs are from manufacturers' catalogs and sales representatives. Labor costs are from R.S. Means [3]. All costs include 1570 overhead and profit. Material and labor costs for the ground-source heat pump excavation and piping are included in the material cost column above.

Of the five options, the gas-fired heat pump was the winning technology (i.e., had the highest NPV) for this ERO. However, because natural gas is not now available at Fort Irwin (and it is unknown if the unit can be converted to LPG), this option is not viable at the Fort. In addition, present or future air quality laws may restrict the use of individual natural gas engines at each housing unit, and the residential gas-fired heat pump technology is still only in the testing stages and is therefore not available.

The runner-up HVAC technology is the high-efficiency ground-source heat pump. Full implementation of this ERO has an initial cost of $7,086,917, with a savings-to-investment ratio of 1.7 and discounted payback of 9.9 years.

It is estimated that the most cost-effective implementation of this ERO will result in an *increase* in annual electric energy consumption of 299,851 kWh but an accompanying decrease in propane use of 77,702 MBtu, for a total annualized energy cost savings of $479,316 and electric demand savings of 15,226 kW-month at an annualized value of $234,468.

The gas-fired heat pump is the only option that would require significant additional maintenance; the oil, oil filter and spark plug must be replaced yearly at an estimated cost (materials and labor) of $105 per unit. Replacing the furnace and air conditioner with a heat pump should result in minor O&M savings of approximately $16,370/yr.

SUMMARY

Potential energy conservation measures, fuel-switching opportunities, and renewable energy projects at facilities the size of Fort Irwin are innumerable. A practical method to systematically assess all possible combinations of energy resource opportunities is needed to make the selection, evaluation, and prioritization of individual energy projects a manageable task. The FEDS process and Level-2 software do just that.

A FEDS assessment of Fort Irwin was recently completed. Significant energy and energy-cost savings opportunities were identified that would reduce building electric energy consumption by 16%

and propane consumption by 34%. For non-building EROs, electric energy consumption would be reduced by 15%; fossil-fuel consumption would decline by 14%.

Individual energy projects identified using the FEDS process have been folded into a 5-year energy plan currently being implemented at Fort Irwin. It is anticipated that the FEDS assessment will be repeated in 1996 to reevaluate the energy conservation opportunities, given implementation of some of the recommended projects, revised energy costs, and other changing conditions at Fort Irwin.

ACKNOWLEDGMENT

The Pacific Northwest Laboratory is operated by Battelle Memorial Institute for the U.S. Department of Energy under Contract DE-AC06-76RL0 1830.

REFERENCES

[1]   10 CFR 436. 1992. U.S. Department of Energy, "Federal Energy Management and Planning Programs." *U.S. Code of Federal Regulations.*

[2]   Richman, E.E., J.M. Keller, A.L. Dittmer, and D.L. Hadley. 1994. *Fort Irwin Integrated Resource Assessment Volume 2: Baseline Detail.* PNL-9064 Vol. 2, Pacific Northwest Laboratory, Richland, Washington.

[3]   Means. 1992. *MEANS Building Construction Cost Data: 1992 15th Annual Edition.* R.S. Means Company, Inc., Kingston, Massachusetts.

13

Case Study:
An Integrated Approach
To Achieving Energy
Cost Savings at
Corporate Headquarters

INTRODUCTION

Over the past 20 years, many companies have demonstrated their commitment to energy conservation as a means of cost reduction. A measure of their success has been substantially reduced energy consumption compared to previous years. Companies have not only been rewarded for controlling energy costs on the bottom line, they also achieved the added benefit of environmental contributions such as reduced direct or indirect air emissions and reduced usage of fuel resources. Indeed, environmental policies are major factors driving today's industrial manufacturing businesses to continue or increase energy management activities.

In contrast to industrial manufacturing and the associated production facilities, are the corporate staff facilities that are still maintained to varying degrees by many major corporations. Such facilities are sometimes geographically distributed or centrally located at a single common site. Business functions including research, product

Presented at 17th World Energy Engineering Congress by David A. Eberly testing, marketing development, engineering, and business information analysis might be performed at these locations. For either type of location, the energy operating costs of the corporate staff facility can be a significant expense. Therefore, all efforts to reduce these costs are desirable. When cost effective activities are identified, they are normally implemented.

An integrated approach to energy cost reduction brings together many conventional and non-conventional energy management strategies. Initial consumption reductions are derived from conventional strategies such as higher thermostat settings in warm weather, lower thermostat settings in cold weather, reduced HVAC operating times, and lighting system improvements. These strategies are obviously important since they represent a significant amount of the potential energy savings. In recent years however, less conventional strategies have been invented. Technological developments have made it possible to gain the benefits of improved efficiency and lower operating costs without the need to compromise environmental working conditions. In addition, natural gas and electric utility suppliers find themselves in a changing, and increasingly, more competitive business environment. As a result, creative rate options and demand-side management programs have been developed to both retain current customers and attract new ones.

This chapter highlights a comprehensive, integrated energy management process implemented at a corporate staff facility over a three-year period. In addition to descriptions of traditional energy efficiency/improvements, this chapter describes the facility complex, baseline energy utilization prior to implementing improvement strategies, and the less traditional methodologies applied.

BACKGROUND

Since the oil embargo in 1974, all United States market segments have been exposed to incentives to identify and implement energy conservation strategies. Back in the '70s and early '80s—the real conservation motivators were the need for adequate supplies of energy as well as cost curtailment Energy management suddenly became part of many businesses strategic planning process. Today, domestic and global competition to be the lowest cost producer is a prime motivator. Environmental factors significantly contribute to our need to be more energy efficient, to cause fewer direct and indirect emissions, and to eliminate all waste.

FACILITY DESCRIPTION

The following describes the conservation and efficiency improvement efforts at Armstrong World Industries' Innovation Center, located in Lancaster, Pennsylvania. While there were one time initiatives to control energy costs prior to 1990, a more structured and continuous effort was formalized in 1990. This process is in place and continues to evolve today.

The Innovation Center's facilities are located on a 600-acre site in rural Lancaster County, Pennsylvania. The area averages more than 6,000 heating degree days annually. There are 19 buildings occupying only a small portion of the physical site—representing 660,000 square feet of conditioned space. The buildings range in age from those initially constructed in 1950 to the latest Building SA addition, which was commissioned in April 1992. Facilities vary in construction techniques, envelope structure, HVAC design, etc. With the exception of Building 19 and the Marketing Development Center, all facilities make use of centralized steam and chilled water plants. Other elements such as lighting, insulation levels, and HVAC design are characteristic of state-of-the-art concepts for the construction time period.

ENERGY SYSTEMS

Utility types and various energy systems of the facility complex are briefly outlined below. When the center was originally constructed in 1951, the concept of providing centralized plant utility systems for all buildings was common practice. Steam used for building heat, domestic hot water, and research pilot manufacturing operations is supplied from a characteristic boiler house. It consists of three water tube boilers—15,000, 15,000, and 40,000 #/hr.—all firing natural gas. Steam is produced at nominal header pressure of 150 psig, with total production varying from 6,400 (summer average) to 28,000 pounds per hour. Summer load consists mostly of air conditioning reheat energy supplied to five buildings.

Chilled water, used exclusively for building comfort air conditioning, is produced by a central chiller plant. It consists of four centrifugal chillers, with total capacity of 2,900 tons. Nominal summer continuous load is 1,400 tons, peaking at 1,800 tons. Worth noting is the fact that these machines are all chlorofluorocarbon (CFC) charged, and therefore, the subject of a current conversion/replacement evaluation.

Compressed air is utilized by research pilot manufacturing processes for process controls, valves, cleaning, and various pneumatically powered equipment. Air is also required by most building HVAC controls, which use air to operate dampers, positioners, valve actuators, and environmental controls. Two reciprocating compressors, with 930 cfm capacity, representing about a 225 horsepower electrical load, produce the air in the boiler room complex.

Finally, electrical power is purchased from the local utility at 13,800 volts and distributed to seven (7) substations. These vary in size from 500 kva to 3750 kva. Secondary power distribution voltages of 4,160 vac, 480, and 240/120 volts are used. Power metering has also been incorporated at some of the substations. Peak electrical power loads occur during the summer months, typically during July or August. In recent years, peak demand has exceeded 4,700 kilowatts. Energy usages also increase during the cooling season, a result of operating the air conditioning systems.

INTEGRATED ENERGY MANAGEMENT

In 1991, Innovation Center facility management came to believe that the trend of rising energy costs could be controlled and reduced with relatively minimal effort. Based on discussions with other local industries and internal discussions at Armstrong, it was determined that there were no existing formal efforts to control energy use or waste at the facility. Since the corporation was already deeply committed to quality management and the team problem solving (participative management) approach to business problem analysis, an energy review committee was appointed in 1991. This group consisted of a representative from each building or technical function, the corporate energy engineer, and a management representative. The purpose of this committee was originally twofold:

- to create awareness of energy use, waste, and costs for all facility work areas

- to develop and implement energy utilization plans that permit the facility to operate with reduced amounts of energy and lower cost

Given these goals, the first activities undertaken by the committee were to develop a plan to reduce energy use for the approaching Christmas holiday period. Methods identified and implemented included simple equipment shutdowns, temperature setbacks, and the activation of some automatic controls previously set for manual operation during these periods. Results of this effort, which covered an eight-day period, were impressive and well received by facility management. Compared to the previous holiday period, 1,500,000 pounds of steam and 50,000 kWh of electricity were conserved, saving more than $10,000 for the eight-day holiday period.

Encouraged by these results, the next period of committee activity involved identifying those procedures or activities that might be cost effectively automated, to take advantage of regular daily and weekend periods. Similar activities that could be performed by central plant utility operators were also evaluated.

Time Clock Applications

The research facility was targeted for simple seven-day time clock application to automatically control building air handlers. A simple time-of-day schedule was implemented for all HVAC fans in Buildings 1, 2, and 3. This timer controls 150 kW of electrical load, saving more than 340,000 kWh the first year of operation. Other existing timers for similar equipment such as rooftop systems, lighting, and hot water systems were activated where feasible.

Air Pressure Reductions

Electrical power testing was conducted on the 530 cfm air compressor which supplies the compressed air requirements during off-hour periods. Reducing setpoint air pressure from 85 psig to 71 psig (minimum to operate regenerative air dryer valves) reduced power consumption by nearly 20 percent. Approximately 71,000 kWh is saved annually, or $3,000.

Lighting

Today, a significant focus area for the corporation is lighting efficiency improvements. It is estimated that lighting accounts for 15 percent of total corporate electricity usage, and more than 35 percent of the power used at the Innovation Center complex. Even before the National Energy Policy Act of 1992 (NEPA) focused national attention on lamp efficiencies, Armstrong was active in the replacement or elimination of standard or non-energy saving lamps. Several energy efficient lighting technologies have been incorporated to reduce lighting power requirements by 30 percent or more in some buildings. These include:

- **Fluorescent lamps** - The standard F40CW lamp was gradually replaced with several energy-saving equivalents by stocking and standardizing in the facility storeroom. Thousands of cool white, warm white, and designer phosphor lamps have replaced the F40CW, saving 6 watts or 120 kWh over the life of the lamp.

- **Compact fluorescent** - In the 80,000 square foot engineering building, 265 recessed incandescent down lights were retrofit with CF lamps. The change from 75 watt incandescent to 18 watt compact fluorescent technology saved 130,000 kWh the first year or $7,300 while providing a 170 percent investment rate of return. CF lamps have also been applied in stairwells and for security lighting in parts of the complex.

- **Halogen incandescent** - While we have strived to eliminate incandescent sources wherever a more efficient lamp will work, there continue to be applications where the old standby is still preferred. These include dimming applications, retail and marketing displays, and color sensitive task and down lighting applications. Halogen A-line, PAR, and reduced voltage MR lamps have been used extensively to meet these requirements while delivering more lumens with fewer watts.

- **T-8 fluorescent** - No T-8 lamp retrofits have been implemented at the facility, primarily because the economics have not met past or current investment hurdle rates. However, this technology has been used in all new installations including the 1992 Building 5A addition. It has also been established as the corporate standard for new fluorescent lighting applications. The technology has also been widely and successfully used at other locations, particularly where utility rebates have contributed to the premium costs for the system.

- **Occupancy sensors** - Motion sensors have been installed for some individual offices, restrooms, utility and conference rooms. While they are usually cost-effective controls in these applications, relatively small amounts of energy are affected. These areas also serve to increase energy awareness, specifically the need to turn off lighting, when practical, in unoccupied areas.

Utility Rate Incentives

Recently, innovative utility rates created incentives for customers to strongly consider their patterns of energy consumption. These incentives typically take the form of either higher or lower rates during designated on-peak periods. The most common rates now available are time-of-use rates (TOU) and interruptible and curtailable rates (I&C), primarily available to industrial customers. With interruptible rates, customers pay reduced demand and energy charges in exchange for an agreement to curtail power use when requested by the utility during designated periods of high demand or system "emergencies." This is a less traditional energy concept when compared to conservation or efficiency improvements.

An interruptible electric rate became available to the Innovation Center in late 1992, and was implemented in January of 1993. The complex contracted for about 35 percent of the previous year's peak summer demand. Curtailment procedures were developed, guided by the efforts of Energy Committee members. The procedures address major electrical loads such as chillers, pumps, building air handlers, and existing standby or emergency generators. Discretionary loads such as general lighting, work area machinery, non-centralized air conditioners, and test equipment are also included in the procedures. Significant savings resulted from the rate change, highlighted in Table 13-1. Per unit electricity costs decreased by approximately 20 percent.

TABLE 13-1. Interruptible Power Rate Comparisons

Rate	Firm kW	Savings	Cost/kWh
Firm	100%	0%	$.061
Interruptible	3,000	3.3%	$.059
	2,500	8.2%	$.056
	2,000	14.7%	$.052
	1,500	19.6%	$0.49
	1,000	26.2%	$.045
	0	36.0%	$ 039

Power Monitoring System

In conjunction with the change to an interruptible power purchase rate, it was recognized that in order to "manage" a facility wide electrical curtailment, a real time power measurement system would be required. Most frequently, such systems are installed to identify where a facility's power is being consumed, wasted, and to more accurately assign costs to manufacturing operations.

At the Innovation Center, such a system was applied to take the "guesswork" out of reaching a target firm power level during curtailment periods. A six-point personal computer based electronic system was installed at the complex, retrofitting existing power system watthour metering. These monitors were invaluable during all four curtailments since installation. They were used to verify that identified equipment loads had in fact been shut off, stayed off, and that the rate of power usage did not exceed utility contract limits. The facility was successful in each case of reaching firm usage levels within the two-hour notification period required by the utility.

Energy Management Control System

After refining shutdown procedures for daily and weekend periods, it became apparent that there was significant opportunity to automate most site building systems. Automatic controls could vary HVAC operation and automatically adjust temperature and run time set points based on time-of-day parameters. An engineering project was funded to replace the existing pneumatic controls in the engineering building with an Energy Management Control System (EMCS) incorporating direct digital controls. Twelve air handlers and seven exhaust systems were incorporated into the design. Equipment costs were totally justified on anticipated savings from optimized HVAC run times, reduced central plant chilled water and steam requirements The system was commissioned in the fall of 1993. Table 13-2 shows the reduction in direct electricity usages, compared to the 1991 base periods. Indirect savings from central plant utilities has been estimated to be two times the direct savings.

Variable Speed Drives

As an element of the EMCS project just described, two variable speed fan drives were incorporated into the largest of seven paired air handlers in the engineering building. The 64,000 cfm supply fan and 38,000 cfm return fan were retrofit with 40 hp and 30 hp drives respectively. The rationale for the installation deals specifically with the type of HVAC system in the building, and the belief that varying ventilation rates as a function of occupancy was a practical energy management technique. A constant volume terminal reheat system is

TABLE 13-2. Innovation Center—Summary Energy Results

MANAGEMENT ACTIVITY	FUEL REDUCTION	POWER REDUCTION	$$ SAVINGS
Holiday Period Shutdowns	1,500 MMBtu	50,000 kWh	$10,000
Timeclock Applications	N/A	340,000 kWh	$17,000
Air Pressure Reductions	N/A	71,000 kWh	$3,000
Lighting Improvements	N/A	354,000 kWh	$18,300
Electric Rate Savings	N/A	N/A	$261,000
EMCS & VSD Applications	3,045 MMBtu	790,000 kWh	$71,700
TOTALS	4,545 MMBtu	1,605,000 kWh	$381,000

utilized in this building. Even though the building is only fully occupied from 8:00 a.m. to 4:30 p.m., Monday through Friday, and 5 percent occupied on weekends, prior to the EMCS, all fans were operated at full speed to midnight daily and from 7:00 am. to 3:30 p.m. on weekends. Given that daily occupancy after 5:00 p.m. consists of a cleaning staff, and perhaps five to seven engineers, building operators believed that ventilation rates could be changed to accommodate the significantly reduced load. Drives were justified based on reducing ventilation rates by 50 percent during these nonoccupied periods.

SUMMARY RESULTS

Table 13-2 summarizes the more significant and measurable activities that have been implemented at the Innovation Center since 1991. Annualized energy and cost savings are tabulated.

For the same time period, Figure 13-3 shows that total energy consumed decreased by 9.3 percent per square foot, while total costs have gone down by 16.1 percent.

Figure 13-3.
Engineering Building Power Annual Usage Comparison

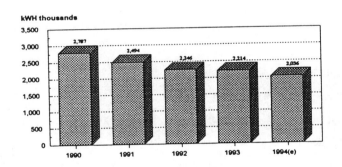

Fans & Lighting Only

FUTURE ACTIVITIES

Considering all the varied projects already undertaken at the Innovation Center, several site opportunities will be evaluated for future funding. These include:

- **EMCS Expansion** - The existing single building system will likely be expanded to cover and control the entire Innovation Center site. The expansion will encompass Buildings 1, 2, and 3, the Product Styling building, and the central chiller plant

- **Monitoring System Expansion** - Power monitors will be added for more refined measurement and electricity use accountability by building.

• **Variable Speed Drives** - Since there are four additional constant volume air handlers in the engineering building, a retrofit project will be considered for these later this year. There are similar opportunities in other buildings, which will likely be evaluated for 1995 budgets.

• **Utility Rate Changes** - Given the success of 1993 power curtailments and the experience of managing each interrupt, facility management will likely decide that an even lower level of firm power usage is appropriate for the site. Again, the table in Figure 13-4 shows the savings potential.

Figure 13-4.
Innovation Center Energy Comparisons

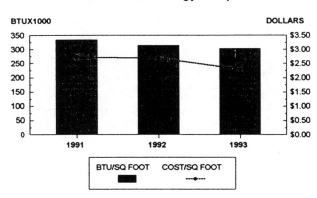

SUMMARY

Armstrong has been proactive in their pursuit of energy management for all their domestic facilities since 1991. From this paper's discussion, it is evident that to manage energy, an integrated approach involving people, equipment improvements, and the ability to take advantage of creative utility rates are required. An awareness of external market factors affecting utility rates, equipment technologies, and the application knowledge are key to implementing this strategy. The energy management process is continuous, multidisciplined, and produces tangible benefits that protect the environment while improving the bottom line.

14

In Transition from Energy Audits to Industrial Assessments

INTRODUCTION

The present Energy Analysis and Diagnostic Centers' program, sponsored by the U.S. Department of Energy is in transition from providing industrial energy audits to providing industrial assessments. This chapter presents the perspective of one of the centers which is currently undergoing this transition. This process of transition includes a change from the point of view of training future energy engineers. Currently, the focus is on energy management engineering for the U.S. manufacturing sector. The new focus will include multi-facet assessments for energy savings, waste minimization, and process Improvements.

The expanded role of the newly-converted centers combines an interdisciplinary approach of engineering analysis which includes the traditional engineering disciplines of electrical, mechanical, chemical, and industrial engineering. These newly-converted centers have the responsibility to perform industrial assessments, hence the new name for these centers is Industrial Assessment Centers indicating the transition from recommending energy conservation opportunities to broader-based assessments.

This chapter will give examples of the expanded scope in assessments for various technologies showing the role of each of the three

Presented at 17th World Energy Engineering Congress by John W. Sheffield and Burns E. Hegler facets, i.e. energy savings, waste minimization and process improvements. The technologies include alternative refrigerants, reclamation & recycling, pollution prevention technologies, natural gas opportunities and new high-efficiency products. Illustrations from future case studies might show some compounding effects for opportunities which were not previously diagnosed as economically feasible, which are now becoming more attractive. The compounding effects are not limited to the interdependence of energy savings and waste minimization but also the improved process gains.

BACKGROUND

The U.S. Department of Energy (DOE), Energy Analysis and Diagnostic Center Program (EADC) has offered no-cost energy conservation audits to industrial plants since 1976. In 1988 the Waste Minimization Assessment Center (WMAC)[1] program funded by the U.S. Environmental Protection Agency was initiated in order to assist small and medium-size manufacturers. An initiative for phasing experienced EADCs into Industrial Assessment Centers (IAC) began in 1993. The establishment of the IACs provides the small and medium-size manufacturers with a combined energy, waste and process assessment. The energy assessment includes identification of energy conservation opportunities, annual cost savings, annual energy savings, implementation costs and paybacks. The waste assessment includes identification of waste minimization opportunities, annual cost savings, annual mass savings, implementation costs and paybacks. Finally, the process assessment includes identification of process efficiency improvements, annual cost savings, implementation costs and paybacks.

As one of the successful programs offered by the U.S. Department of Energy, Office of Industrial Technologies[2], the EADC program is designed to show how to conserve energy and reduce costs for small and medium-sized manufacturers. Engineering faculty and students, who perform these no-cost energy audits, identify all energy using systems in the plant. After the visit, the EADC issues a detailed, confidential report outlining the manufacturing plant's present energy usage and presenting specific energy conservation opportunities (ECOs). Each ECO includes appropriate technology and economic justification. A total of thirty energy audits per year are performed by each EADC.

ENERGY CONSERVATION OPPORTUNITIES

One approach to categorizing the various ECOs originates from the energy source, i.e. electricity, fossil fuels, and alternate energy sources. The fossil fuels can be subdivided into natural gas, propane, fuel oils, and coal. Examples of alternative energy sources include wood and waste materials.

Electricity-Sourced ECOs

Within the various electricity-sourced ECOs, one can categorize these ECOs by the end-use equipment or function, such as electric motors, lighting systems, air compressors, cooling towers, chillers, electric water heaters, electric ovens, electric furnaces, refrigeration units, electric space heaters, transformers, fans, blowers, and other electric devices. For each type of end-use equipment or function, a series of energy conservation opportunities exist. Table 14-1 gives a set of potential ECOs that might be considered for lighting systems at a manufacturing plant.

TABLE 14-1. Lighting System ECOs

Install High-Frequency Electronic Ballasts
Install High Efficacy Lamps
Install Occupancy Sensors
Install Photosensors & Utilize Daylighting
Reduce Lighting Usage

These lighting ECOs illustrate the potential for new technologies to be adopted into practice based on cost savings due to improved energy efficiency. Other electric sourced ECOs represent the energy conservation practice in good engineering, such as in air compressors. Table 14-2 illustrates a series of ECOs for air compressors.

TABLE 14-2. Air Compressor ECOs

Use Synthetic Lubricants in Air Compressors
Reduce Compressed Air Pressure
Recover Compressor Waste Heat
Reduce Compressed Air Leaks
Install Larger Header Line in Compressed Air System
Use Outside Air for Compressor Intake

Other electricity-sourced ECOs reflect potential trade-offs between initial costs, operating cost associated with energy, and cycle life of the equipment. Electric motors and their associated mechanical drives have several ECOs. Table 14-3 lists a set of potential ECOs for electric motors.

TABLE 14-3. Electric Motor ECOs

Install High-Efficiency Motors
Install Variable Speed Controls
Replace Standard V-Belts with High-Efficiency Belts
Install Synchronous Belts & Drives

In general each type of electricity-sourced end-use equipment would have a corresponding set of potential energy saving opportunities.

Natural Gas ECOs

Natural gas energy conservation opportunities can be sorted by end-use equipment such as boilers, burners, ovens, furnaces, coolers, and heaters. In addition, several cost saving opportunities exists without a corresponding energy savings. An example of these cost savings could be the purchase of contract natural gas to achieve lower unit cost of the fuel. Table 14-4 gives a set of potential ECOs for a natural gas fired boiler.

TABLE 14-4. Boiler ECOs

Improve Boiler Combustion Efficiency
Install Condensate Return Systems
Install High-Pressure Condensate Return Systems
Install Tubulators in Boiler Tubes
Install Small Boilers
Recover Steam Blowdown
Repair Steam Traps
Install Feedwater-Preheater Systems
Shut Off Boiler during Idle Periods
Duct Warm Combustion Air to Boilers
Repair Condensate Leaks
Repair Steam Leaks

As a point of reference, the first ECO listed in Table 14-4 entitled *Improve Boiler Combustion Efficiency* might involve several recommended actions. For example, a typical recommendation for a natural gas fired boiler would be to clean and adjust the air-to-fuel ratio to achieve an improved combustion efficiency. However, some boilers might be candidates for the installation of an oxygen sensor combined with a continuous trimming of the combustion air to achieve an optimum air-to-fuel ratio under all operating conditions and variations of natural gas composition.

General ECOs

There exists many end-use devices that are general in that their function is not dependent of the energy source. Examples of such equipment include heat exchangers, thermal insulation, stack dampers, infiltration inhibitors, heat recovery, and various controllers. Table 14-5 lists a set of general ECOs that might be examined for potential applicability at each manufacturing plant for cost savings.

TABLE 14-5. General ECOs

Install Stack Heat Exchangers
Install Covers for Heated Tanks
Install Dock Seals or/and Dock Shelters
Install Strip Doors (Interior/Exterior)
Install Stack Dampers
Install Destratification Fans
Install Radiant Heaters
Install Automatic Clock Thermostats
Install Energy Management Systems
Install Exhaust Hood for Ovens
Insulate Pipes/Ovens/Boilers/Dock Doors/Ducts/etc.
Balance Make-Up Air Systems

WASTE MINIMIZATION OPPORTUNITIES

With the increasing cost of management and disposal waste material, including process-related and residues from waste treatment, for manufacturers, a logical approach to minimizing the effect and stress on the environment is to reduce or eliminate the waste as its source. In the past, the WMAC identified and analyzed waste minimization opportunities (WMO). Specific WMO were recommended and the essential supporting technological and economical information was developed and presented to the manufacturing clients in the form of waste minimization assessment reports at a no out-of-pocket cost.

The classification of WMOs can be divided into source reduction, material substitution, recycling, waste treatment, and alternative waste management techniques. One can draw analogies between each of the basic classes of WMOs and a corresponding classes of ECOs. For example, a source reduction WMO might have a parallel energy conservation opportunity due to reduced material handling. Likewise a material substitution WMO might have a corresponding ECO related to a change in energy source, i.e. higher unit energy cost of electricity to a lower unit energy cost of natural gas. For some recycling WMOs, for example, one might be able to recover a solvent as a fuel additive ECO.

Source Reduction WMOs

The source reduction WMOs would be the natural first choice for a manufacturer. The direct cost savings would be clearly identified and accountable. Table 14-6 gives several candidate source reduction WMOs.

TABLE 14-6. Source Reduction WMOs

Reduction of Liquid *Drag-Out*
Reduction of Solid *Drag-Out*
Reduction of Water Use
High Transfer Efficiency Spray Paint Guns

Material Substitution WMOs

The material substitution WMOs might be especially recommended when one could substitute a non-hazardous substance for a hazardous substance. Table 14-7 presents several types of material substitution WMOs.

TABLE 14-7. Material Substitution WMOs

Alternative Cleaners or Solvents
Alternative Cleaning Methods
Other Material Substitutions

Recycling WMOs

Recycling WMOs receive significant attention because of various local, state and federal regulations. In addition, the economics of recycling can be significant for manufacturers. In pursuing recycling goals, manufacturers might well develop an awareness program. For example, they might develop a program of rational metal-working oils and coolants management. The positive benefit of having such an awareness program among the labor forces can be well recognized by the local, state, federal and even world community. While the environmental issues facing the manufacturing industry today have expanded considerably beyond the traditional concerns of recycling, these WMOs are still viable. Table 14-8 gives several candidate Recycling Waste Minimization Opportunities specifically suitable to the manufacturing industry.

Treatment WMOs

The assessment process to identify WMOs requires consideration of the manufacturer's process operation, basic chemistry, and various environmental concerns and needs. The manufacturers have become increasingly concerned with wastewater treatment, air emissions, potential soil and groundwater contamination, solid waste

TABLE 14-8. Recycling WMOs

Solvent Recovery and Reuse
Solvent Recovery and Distillation
Filtration and Agitation
Water Reuse
Off-site Recycling
Metal-Cutting Fluid Management

disposal and employee health and safety. For example in the micro-electronics industry, potential WMOs include the evaporation of sodium hydroxide waste and the use of ion exchange systems.

Alternative Waste Management WMOs

Various waste segregation or exchange opportunities might be considered as alternative waste management WMOs. For example, a more thorough segregation of scrap materials such as plastics, metals, and wood products might be a candidate WMO. Another example of an alternative waste management WMO is the segregation of waste solvents and sludge. Yet another example of an alternative waste management WMO is the de-emulsification and segregation of waste oils.

PROCESS IMPROVEMENT OPPORTUNITIES

The recommendations from the new IACs will include not only energy conservation opportunities and waste minimization opportunities, but also process improvement opportunities (PIO). With continuous improvement being a key to competing in the 90's, manufacturers need an assessment of how efficient their plant facilities are and how to improve their processes. They would like to know how to improve their profits while maintaining customer satisfaction. Implementation of PIOs depend on the manufacturer's goals, constraints, budgets, and time frame. Of course, each manufacturer would like to know how their plant ranks with their competitors on manufacturing performance. Recommended PIOs by an IAC can provide a manufacturer with access to alternative technologies to improve their efficiency. In addition, these PIOs can help correct problems causing low quality products, low productivity or low morale. In the past, the recommendations made by either the EADCs or the WMACs provided their clients with a list of recommendations that established a plan for improved energy efficiency or waste minimization management. Now the IACs have the benefit of both of those along with a plan for continuous improvement in manufacturing processes.

SUMMARY

The capabilities of the Industrial Assessment Centers in combining energy conservation, waste minimization and process improvement recommendations is not unique for small and medium-sized manufacturers. For example, the Mid-America Manufacturing Technology Center (MAMTC)[3] is a non-profit organization designed to improve the competitiveness and productivity of small and medium sized manufacturers. However, one difference between the IAC program and the MAMTC program is that the IAC operates on a *no out-of-pocket cost* basis.

In conclusion, the goals of any such program to improve the competitiveness and productivity of our small and medium sized manufacturers includes the delivery of a timely evaluation report that covers: technology, operation, quality, and safety issues. An integral part of such a report would be the inclusion of a combined energy, waste and process efficiency assessment.

REFERENCES
[1]   "Waste Minimization Assessment Center - An EPA Program for Small and Medium-Size Manufacturers," Industrial Technology and Energy Management Division, University City Science Center, Philadelphia, PA.

[2]   "The Office of Energy Efficiency and Renewable Energy,"
      DOE/CH10093-160, June 1993.

[3]   CITE (Continuous Improvement Targets for Excellence) As-
      sessment Program, NIST/Mid-America Manufacturing Tech-
      nology Center, Overland Park, KS, 1994.

15

A Compendium Of Handy Working Aids

This chapter contains tables, figures, and forms to supplement information in the foregoing chapters. Examples of energy audit forms are presented. Feel free to modify these forms to meet your requirements.

Table 15-1. Degree Day Data

(Source: Cooling and Heating Load Calculation Manual ASHRAE GRP 158)

Average Winter Temperature and Yearly Degree Days for Cities in the United States and Canada[a,b,c] (Base 65°F)

State	Station		Avg. Winter Temp.[d] F	Degree-Days Yearly Total	State	Station		Avg. Winter Temp. F	Degree-Days Yearly Total
Ala.	Birmingham	A	54.2	2551	Calif.	Bakersfield	A	55.4	2122
	Huntsville	A	51.3	3070		Bishop	A	46.0	4275
	Mobile	A	59.9	1560		Blue Canyon	A	42.2	5596
	Montgomery	A	55.4	2291		Burbank	A	58.6	1646
Alaska	Anchorage	A	23.0	10864		Eureka	C	49.9	4643
	Fairbanks	A	6.7	14279		Fresno	A	53.3	2611
	Juneau	A	32.1	9075		Long Beach	A	57.8	1803
	Nome	A	13.1	14171		Los Angeles	A	57.4	2061
Ariz.	Flagstaff	A	35.6	7152		Los Angeles	C	60.3	1349
	Phoenix	A	58.5	1765		Mt. Shasta	C	41.2	5722
	Tucson	A	58.1	1800		Oakland	A	53.5	2870
	Winslow	A	43.0	4782		Red Bluff	A	53.8	2515
	Yuma	A	64.2	974		Sacramento	A	53.9	2502
Ark.	Fort Smith	A	50.3	3292		Sacramento	C	54.4	2419
	Little Rock	A	50.5	3219		Sandberg	C	46.8	4209
	Texarkana	A	54.2	2533					

[a] Data for United States cities from a publication of the United States Weather Bureau. *Monthly Normals of Temperature, Precipitation and Heating Degree Days*, 1962, are for the period 1931 to 1960 inclusive. These data also include information from the 1963 revisions to this publication, where available.

[b] Data for airport station, A, and city stations, C, are both given where available.

[c] Data for Canadian cities were computed by the Climatology Division. Department of Transport from normal monthly mean temperatures, and the monthly values of heating days data were obtained using the National Research Council computer and a method devised by H. C. S. Thom of the United States Weather Bureau. The heating days are based on the period from 1931 to 1960.

[d] For period October to April, inclusive.

Table 15-14. Degree Day Data (con't)

State	Station		Avg. Winter Temp. F	Degree-Days Yearly Total
Calif. (Con'td)	San Diego	A	59.5	1458
	San Francisco	A	53.4	3015
	San Francisco	C	55.1	3001
	Santa Maria	A	54.3	2967
Colo.	Alamosa	A	29.7	8529
	Colorado Springs	A	37.3	6423
	Denver	A	37.6	6283
	Denver	C	40.8	5524
	Grand Junction	A	39.?	5641
	Pueblo	A	40.4	5462
Conn.	Bridgeport	A	39.9	5617
	Hartford	A	37.3	6235
	New Haven	A	39.0	5897
Del.	Wilmington	A	42.5	4930
D.C.	Washington	A	45.7	4224
Fla.	Apalachicola	C	61.2	1308
	Daytona Beach	A	64.5	879
	Fort Myers	A	68.6	442
	Jacksonville	A	61.9	1239
	Key West	A	73.1	108
	Lakeland	C	66.7	661
	Miami	A	71.1	214
	Miami Beach	C	72.5	141
	Orlando	A	65.7	766
	Pensacola	A	60.4	1463

State	Station		Avg. Winter Temp. F	Degree-Days Yearly Total
Iowa	Burlington	A	37.6	6114
	Des Moines	A	35.5	6588
	Dubuque	A	32.7	7376
	Sioux City	A	34.0	6951
	Waterloo	A	32.6	7320
Kans.	Concordia	A	40.4	5479
	Dodge City	A	42.5	4986
	Goodland	A	37.8	6141
	Topeka	A	41.7	5182
	Wichita	A	44.2	4620
Ky.	Covington	A	41.4	5265
	Lexington	A	43.8	4683
	Louisville	A	44.0	4660
La.	Alexandria	A	57.5	1921
	Baton Rouge	A	59.8	1560
	Lake Charles	A	60.5	1459
	New Orleans	A	61.0	1385
	New Orleans	C	61.8	1254
	Shreveport	A	56.2	2184
Me.	Caribou	A	24.4	9767
	Portland	A	33.0	7511
Md.	Baltimore	A	43.7	4654
	Baltimore	C	46.2	4111
	Frederick	A	42.0	5087

State	City			
	Tallahassee	A	60.1	1485
	Tampa	A	66.4	683
	West Palm Beach	A	68.4	253
Ga.	Athens	A	51.8	2929
	Atlanta	A	51.7	2961
	Augusta	A	54.5	2397
	Columbus	A	54.8	2383
	Macon	A	56.2	2136
	Rome	A	49.9	3326
	Savannah	A	57.8	1819
	Thomasville	C	60.0	1529
Hawaii	Lihue	A	72.7	0
	Honolulu	A	74.2	0
	Hilo	A	71.9	0
Idaho	Boise	A	39.7	5809
	Lewiston	A	41.0	5542
	Pocatello	A	34.8	7033
Ill.	Cairo	C	47.9	3821
	Chicago(O'Hare)	A	35.8	6639
	Chicago(Midway)	A	37.5	6155
	Chicago	C	38.9	5882
	Moline	A	36.4	6408
	Peoria	A	38.1	6025
	Rockford	A	34.8	6830
	Springfield	A	40.6	5429
Ind.	Evansville	A	45.0	4435
	Fort Wayne	A	37.3	6205
	Indianapolis	A	39.6	5699
	South Bend	A	36.6	6439

State	City			
Mass.	Boston	A	40.0	5634
	Nantucket	A	40.2	5891
	Pittsfield	A	32.6	7578
	Worcester	A	34.7	6969
Mich.	Alpena	A	29.7	8506
	Detroit(City)	A	37.2	6232
	Detroit(Wayne)	A	37.1	6293
	Detroit(Willow Run)	A	37.2	6258
	Escanaba	C	29.6	8481
	Flint	A	33.1	7377
	Grand Rapids	A	34.9	6894
	Lansing	A	34.8	6909
	Marquette	C	30.2	8393
	Muskegon	A	36.0	6696
	Sault Ste. Marie	A	27.7	9048
Minn.	Duluth	A	23.4	10000
	Minneapolis	A	28.3	8382
	Rochester	A	28.8	8295
Miss.	Jackson	A	55.7	2239
	Meridian	A	55.4	2289
	Vicksburg	C	56.9	2041
Mo.	Columbia	A	42.3	5046
	Kansas City	A	43.9	4711
	St. Joseph	A	40.3	5484
	St. Louis	A	43.1	4900
	St. Louis	C	44.8	4484
	Springfield	A	44.5	4900
Mont.	Billings	A	34.5	7049
	Glasgow	A	26.4	8996
	Great Falls	A	32.8	7750

Table 15-14. Degree Day Data (con't)

State	Station		Avg. Winter Temp. F	Degree-Days Yearly Total
Mont. (Con'td)	Havre	A	28.1	8700
	Havre	C	29.8	8182
	Helena	A	31.1	8129
	Kalispell	A	31.4	8191
	Miles City	A	31.2	7723
	Missoula	A	31.5	8125
Neb.	Grand Island	A	36.0	6530
	Lincoln	C	38.8	5864
	Norfolk	A	34.0	6979
	North Platte	A	35.5	6684
	Omaha	A	35.6	6612
	Scottsbluff	A	35.9	6673
	Valentine	A	32.6	7425
Nev.	Elko	A	34.0	7433
	Ely	A	33.1	7733
	Las Vegas	A	53.3	2709
	Reno	A	39.3	6332
	Winnemucca	A	36.7	6761
N.H.	Concord	A	33.0	7383
	Mt. Washington Obsv.		15.2	13817
N.J.	Atlantic City	A	43.2	4812
	Newark	A	42.8	4589
	Trenton	C	42.4	4980

State	Station		Avg. Winter Temp. F	Degree-Days Yearly Total
	Columbus	A	39.7	5660
	Columbus	C	41.5	5211
	Dayton	A	39.8	5622
	Mansfield	A	36.9	6403
	Sandusky	C	39.1	5796
	Toledo	A	36.4	6494
	Youngstown	A	36.8	6417
Okla.	Oklahoma City	A	48.3	3725
	Tulsa	A	47.7	3860
Ore.	Astoria	A	45.6	5186
	Burns	C	35.9	6957
	Eugene	A	45.6	4726
	Meacham	A	34.2	7874
	Medford	A	43.2	5008
	Pendleton	A	42.6	5127
	Portland	A	45.6	4635
	Portland	C	47.4	4109
	Roseburg	A	46.3	4491
	Salem	A	45.4	4754
Pa.	Allentown	A	38.9	5810
	Erie	A	36.8	6451
	Harrisburg	A	41.2	5251
	Philadelphia	A	41.8	5144
	Philadelphia	C	44.5	4486

State	City			
N.M.	Albuquerque	A	45.0	4348
	Clayton	A	42.0	5158
	Raton	A	38.1	6228
	Roswell	A	47.5	3793
	Silver City	A	48.0	3705
N.Y.	Albany	A	34.6	6875
	Albany	C	37.2	6201
	Binghamton	A	33.9	7286
	Binghamton	C	36.6	6451
	Buffalo	A	34.5	7062
	New York (Cent. Park)	C	42.8	4871
	New York (LaGuardia)	A	43.1	4811
	New York (Kennedy)	A	41.4	5219
	Rochester	A	35.4	6748
	Schenectady	C	35.4	6650
	Syracuse	A	35.2	6756
N. C.	Asheville	C	46.7	4042
	Cape Hatteras		53.3	2612
	Charlotte	A	50.4	3191
	Greensboro	A	47.5	3805
	Raleigh	A	49.4	3393
	Wilmington	A	54.6	2347
	Winston-Salem	A	48.4	3595
N. D.	Bismarck	A	26.6	8851
	Devils Lake	C	22.4	9901
	Fargo	A	24.8	9226
	Williston	A	25.2	9243
Ohio	Akron-Canton	A	38.1	6037
	Cincinnati	C	45.1	4410
	Cleveland	A	37.2	6351
	Pittsburgh	A	38.4	5987
	Pittsburgh	C	42.2	5053
	Reading	C	42.4	4945
	Scranton	A	37.2	6254
	Williamsport	A	38.5	5934
R. I.	Block Island	A	40.1	5804
	Providence	A	38.8	5954
S. C.	Charleston	A	56.4	2033
	Charleston	C	57.9	1794
	Columbia	A	54.0	2484
	Florence	A	54.5	2387
	Greenville-Spartenburg	A	51.6	2980
S. D.	Huron	A	28.8	8223
	Rapid City	A	33.4	7345
	Sioux Falls	A	30.6	7839
Tenn.	Bristol	A	46.2	4143
	Chattanooga	A	50.3	3254
	Knoxville	A	49.2	3494
	Memphis	A	50.5	3232
	Memphis	C	51.6	3015
	Nashville	A	48.9	3578
	Oak Ridge	C	47.7	3817
Tex.	Abilene	A	53.9	2624
	Amarillo	A	47.0	3985
	Austin	A	59.1	1711
	Brownsville	A	67.7	600
	Corpus Christi	A	64.6	914
	Dallas	A	55.3	2363
	El Paso	A	52.9	2700

Table 15-14. Degree Day Data (concluded)

State	Station	Avg. Winter Temp. F	Degree-Days Yearly Total		Prov.	Station	Avg. Winter Temp. F	Degree-Days Yearly Total
Texas (Con'td)	Fort Worth	A	55.1	2405	Alta.	Banff	C	—	10551
	Galveston	A	62.2	1274		Calgary	A	—	9703
	Galveston	C	62.0	1235		Edmonton	A	—	10268
	Houston	A	61.0	1396		Lethbridge	A	—	8644
	Houston	C	62.0	1278	B. C.	Kamloops	A	—	6799
	Laredo	A	66.0	797		Prince George*	A	—	9755
	Lubbock	A	48.8	3578		Prince Rupert	C	—	7029
	Midland	A	53.8	2591		Vancouver*	A	—	5515
	Port Arthur	A	60.5	1447		Victoria*	A	—	5699
	San Angelo	A	56.0	2255		Victoria	C	—	5579
	San Antonio	A	60.1	1546	Man.	Brandon*	A	—	11036
	Victoria	A	62.7	1173		Churchill	A	—	16728
	Waco	A	57.2	2030		The Pas	C	—	12281
	Wichita Falls	A	53.0	2832		Winnipeg	A	—	10679
Utah	Milford	A	36.5	6497	N. B.	Fredericton*	A	—	8671
	Salt Lake City	A	38.4	6052		Moncton	C	—	8727
	Wendover	A	39.1	5778		St. John	C	—	8219
Vt.	Burlington	A	29.4	8269	Nfld.	Argentia	A	—	8440
Va.	Cape Henry	C	50.0	3279		Corner Brook	C	—	8978
	Lynchburg	A	46.0	4166		Gander*	A	—	9254
	Norfolk	A	49.2	3421		Goose*	A	—	11887
	Richmond	A	47.3	3865		St. John's*	A	—	8991
	Roanoke	A	46.1	4150	N. W. T.	Aklavik	C	—	18017
Wash.	Olympia	A	44.2	5236		Fort Norman	C	—	16109
	Seattle-Tacoma	A	44.2	5145		Resolution Island	C	—	16021

State	City			
	Seattle	C	46.9	4424
	Spokane	A	36.5	6655
	Walla Walla	C	43.8	4805
	Yakima	A	39.1	5941
W. Va.	Charleston	A	44.8	4476
	Elkins	A	40.1	5675
	Huntington	A	45.0	4446
	Parkersburg	C	43.5	4754
Wisc.	Green Bay	A	30.3	8029
	La Crosse	A	31.5	7589
	Madison	A	30.9	7863
	Milwaukee	A	32.6	7635
Wyo.	Casper	A	33.4	7410
	Cheyenne	A	34.2	7381
	Lander	A	31.4	7870
	Sheridan	A	32.5	7680

Prov.	City		
N. S.	Halifax	C	7361
	Sydney	A	8049
	Yarmouth	A	7340
Ont.	Cochrane	C	11412
	Fort William	A	10405
	Kapuskasing	C	11572
	Kitchner	C	7566
	London	A	7349
	North Bay	C	9219
	Ottawa	C	8735
	Toronto	C	6827
P.E.I.	Charlottetown	C	8164
	Summerside	C	8488
Que.	Arvida	C	10528
	Montreal*	A	8203
	Montreal	C	7899
	Quebec*	A	9372
	Quebec	C	8937
Sasks	Prince Albert	A	11630
	Regina	A	10806
	Saskatoon	C	10870
Y. T.	Dawson	C	15067
	Mayo Landing	C	14454

* The data for these normals were from the full ten-year period 1951-1960, adjusted to the standard normal period 1931-1960.

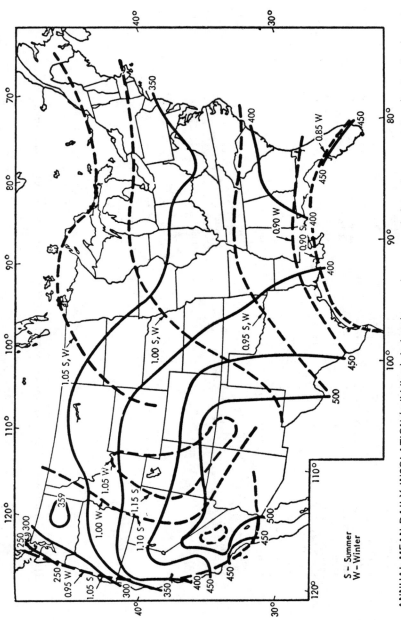

ANNUAL MEAN DAILY INSOLATION (solid lines), in Langleys, and summer and winter clearness numbers (broken lines) are plotted on United States map. Note: To convert Langleys per day to Btu/ft2 day multiply number in figure by 3.69.

Figure 15-1. Annual Mean Daily Insolation

(Courtesy of Heating/Piping/Air Conditioning, Sept. 1966)

S – Summer
W – Winter

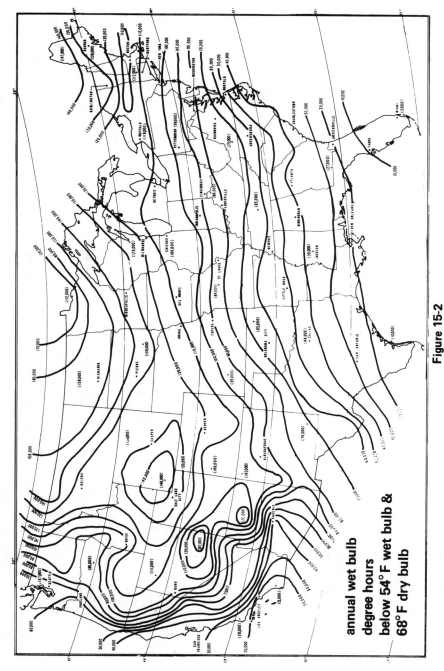

annual wet bulb
degree hours
below 54°F wet bulb &
68°F dry bulb

Figure 15-2

(Source: AFM 88-8, U.S. Government Printing Office, June 15, 1967)

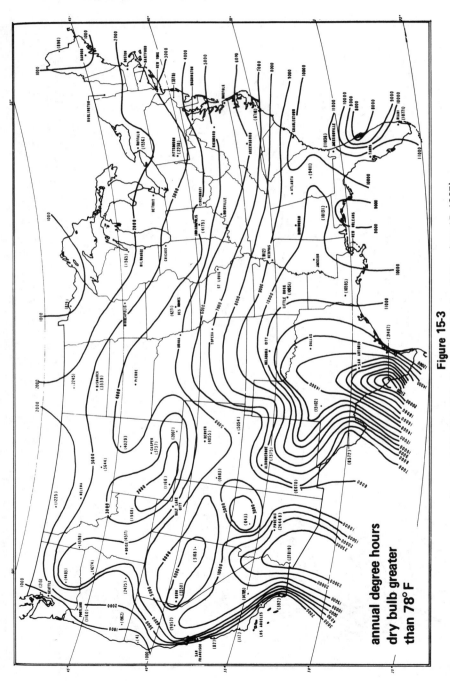

annual degree hours
dry bulb greater
than 78° F

Figure 15-3

(Source: AFM 88-8, U.S. Government Printing Office, June 15, 1967)

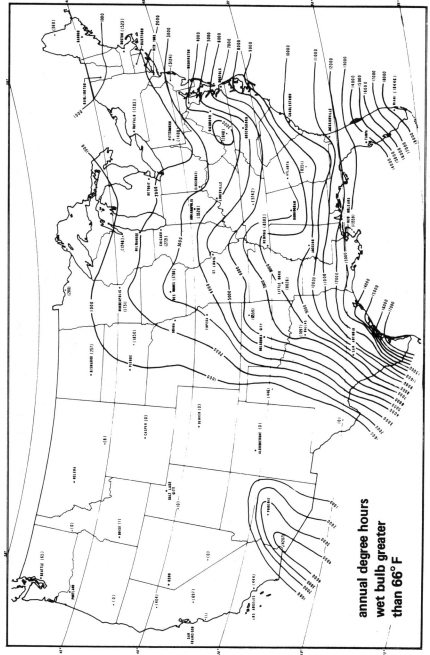

Figure 15-4

(Source: AFM 88-8, U.S. Government Printing Office, June 15, 1967)

annual degree hours
wet bulb greater
than 66° F

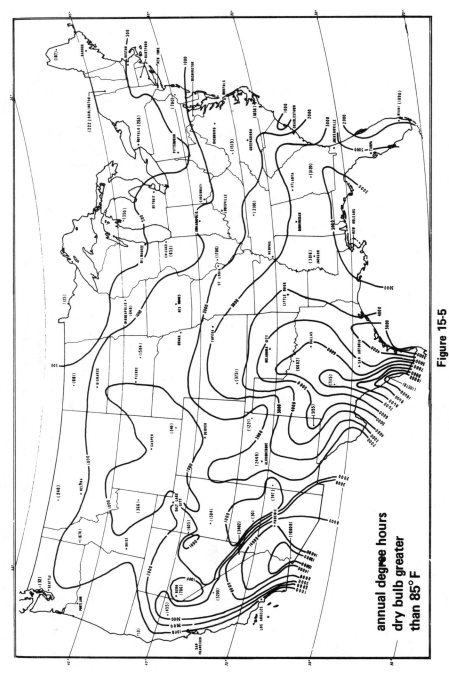

annual degree hours
dry bulb greater
than 85°F

Figure 15-5

(Source: AFM 88-8, U.S. Government Printing Office, June 15, 1967)

BUILDING _____ YEAR _____

MONTH*	HEATING DEGREE DAYS	COOLING DEGREE DAYS	ELECTRICITY QUANTITY KWH	ELECTRICITY COST (DOLLARS) TOTAL $	$/KWH	$/MMBTU	OIL QUANTITY GALLONS	OIL COST (DOLLARS) $/GAL.	TOTAL $	$/MMBTU	NATURAL GAS QUANTITY MCF	NATURAL GAS COST (DOLLARS) TOTAL $	$/MCF	$/MMBTU	COAL☐ WOOD☐ QUANTITY UNIT	PURCHASED STEAM☐ OTHER☐ COST (DOLLARS) TOTAL $	$/UNIT	$/MMBTU	TOTAL ENERGY COST
1	2	3	4	5	6	7	8	9	10	11	12	13	14	15	16	17	18	19	20
JANUARY																			
FEBRUARY																			
MARCH																			
APRIL																			
MAY																			
JUNE																			
JULY																			
AUGUST																			
SEPTEMBER																			
OCTOBER																			
NOVEMBER																			
DECEMBER																			
ANNUAL TOTALS																			
ANNUAL AVERAGES																			

* Or comparable time period

ELECTRICITY = 3412 Btu/kwh
GAS = 1030 Btu/CF

OIL: #2 = .139 MMBTU/gal
     #4 = .150 MMBTU/gal
     #5 = .152 MMBTU/gal
     #6 = .153 MMBTU/gal

MCF = 1000 cubic feet of gas
MMBTU = one million Btu

Figure 15-6. Energy Management Form

1.   Gross Annual Fuel and Energy Consumption

Line No.

	A	B	C
		Conversion Factor	Thousands of BTUs/yr
		x 138 (1) =	
		x 146 (2) =	
1. Oil—gallons			
		x 1.0 (3) =	
2. Gas—Cubic Feet		x 0.8 (4) =	
3. Coal—Short tons		x 26,000 =	
4. Steam-Pounds x 10³		x 900 =	
5. Propane Gas—lbs		x 21.5 =	
6. Electricity—KW.Hrs		x 3.413 =	

7.   Total BTUs X 10³/yr . . . .

8.   BTUs x 10³/Yr/Per Square Foot of Floor Area
(Line 7 ÷ Figure 4, Line 7)
Use for (1) No. 2 Oil; (2) No. 6 Oil; (3) Natural Gas; (4) Mfg. Gas

2.   Annual Fuel and Energy Consumption for Heating

Line No.

	A	B	C
		Conversion Factor	Thousands of BTUs/yr
		x 138 (1) =	
9. Oil—gallons		x 146 (2) =	
		x 1.0 (3) =	
10. Gas—Cubic Feet		x 0.8 (4) =	
11. Coal—Short tons		x 26,000 =	
12. Steam—Pounds x 10³		x 900 =	
13. Propane Gas—lbs		x 21.5 =	
14. Electricity—KW.Hrs		x 3.413 =	

15.   Total BTUs x . . . . =

16.   BTUs x 10³/Yr Per Square Foot of Floor Area
(Line 15   Line 7)

3.   Annual Fuel and Energy Consumption for Domestic Hot Water

Line No.

	A	B	C
		Conversion Factor	Thousands of BTUs/Yr
17. Oil—Gallons		x 138 (1) =	
		x 146 (2) =	

Figure 15-7. Energy Use Audit Form
(Source: Guidelines for Saving Energy in Existing Buildings—Building Owners and
Operators Manual, ECM-1)

	A	B Conversion Factor	C Thousands of BTUs/Yr
18. Gas– Cubic Feet		x 1.0 (3) =	
		x 0.8 (4) =	
19. Coal- Short Tons		x 26,000 =	
20. Steam-Pounds x 10³		x 900 =	
21. Propane Gas- lbs		x 21.5 =	
22. Electricity KW.Hrs		x 3.413 =	
23. Total BTUs/Yr x 10³			
24. BTUs x 10³Yr/Per Square Foot of Floor Area (Line 23 + Figure 4, Line 7)			

4. Annual Fuel and/or Energy Consumption for Cooling (Compressors & Chillers)

Line No.	A	B Conversion Factor	C Thousands of BTUs/Yr
a) if absorption cooling		x 138 (1) =	
25. Oil Gallons		x 146 (2) =	
26. Gas- Cubic Feet		x 1.0 (3) =	
27. Coal Short Tons		x 0.8 (4) =	
28. Steam-Pounds x 10³		x 26,000 =	
29. Propane Gas- lbs		x 900 =	
		x 21.5 =	
30. Total BTUs/yr x 10³			
31. BTUs x 10³/Yr Per Square Foot of Floor Area (Line 30 + Figure 4, Line 7)			

b) if electric cooling

	A	B Conversion Factor	C
32. Electricity- KWH		x 3.413 =	
33. BTUs x 10³/Yr Per Square Foot of Floor Area (Line 32 + Figure 4, Line 7)			

5. Estimated Annual Energy Consumption for Interior Lighting

Line No.	A	B Conversion Factor	C Thousands of BTUs/Yr
34. KWH Fig. 10, Line 3 x Fig. 10, Line 33 (1)		x 3.413 =	
35. BTUs x 10³/Yr/Per Square Foot of Floor Area (Fig. 10, Line 35, Col. C + Fig. 4, Line 7)			

*6. Estimated Annual Electrical Energy Consumption for all Motors and Machines
if Building and Hot Water are Not Electrically Heated: (1)*

36. Total KW Hrs _____ Less KW Hrs Lighting _____ = _____ KW Hrs
(Line 22, Col. A)

37. KW Hrs/Yr/Sq Ft floor area = _____ (1)
(Line 37, Col. C + Fig. 4, Line 7)

38. BTUs x 10³/Yr/Sq Ft floor area = (Line 37) x 3.431 _____ (2)

(1) and (2). If building heat and hot water are electrically heated, deduct the KW
Hrs/Yr per sq ft and BTUs/Yr per sq ft for heating and hot water. (Lines 37 and 38)

Figure 15-7. Energy Use Audit Form (concluded)

PROJECT NO. _____

Job Name _____

System No. _____ Type _____ O.A.T. _____ Time of Day _____ Date _____

Location _____ Tested By _____

1. DRIVE INFORMATION

Motor Manufacturer · · · · · · · · · · _____, Frame Size _____

Motor HP · · · · · · · · · · · · · · · · · _____

Phases · · · · · · · · · · · · · · · · · · · _____

Amperage · · · · · · · · · · · · · · Rated _____ Actual _____

Voltage · · · · · · · · · · · · · · · · Rated _____ Actual _____

Fan RPM · · · · · · · · · · · · · · Regular _____ Actual _____

Fan Manufacturer · · · · · · · · · · · · _____

Fan Type · · · · · · · · · · · · · · · · · · _____

Motor Sheave Position, Type, and Size · · · · · · · · · · · _____

Shaft Diameter · · · · · · · · · · · · · · _____

Key Size · · · · · · · · · · · · · · · · · · · _____

NOTE: ALL TEMPERATURES MUST BE TAKEN AT THE SAME TIME: AND TIME OF DAY, WHEN THE READINGS ARE TAKEN MUST BE INDICATED.

2. FAN DATA

Does system have return fam? · Yes _____ No _____

If Yes, Fan No. · · · · · · · · · · · · · _____

CFM (Design) · · · · · · · · · · Supply _____ Return _____ O.A. _____

CFM (Actual) · · · · · · · · · · Supply _____ Return _____ O.A. _____ at _____ ΔP (inches H$_2$O)

SP Filters · · · · · · · · · · · · Inlet _____ Discharge _____

SP PH Coil · · · · · · · · · · · · Inlet _____ Discharge _____

SP H Coil . . . . Inlet _____ Discharge _____
SP C Coil . . . . Inlet _____ Discharge _____
SP Sup. Fan . . . . Inlet _____ Discharge _____
SP Ret. Fan . . . . Inlet _____ Discharge _____
Temp. Readings . . . . RAT _____ MAT _____ PHDT _____
                                 HCDT _____ CCDT _____ RFDT _____ SFDT _____

3. COIL DATA
Preheat Coil . . . . EWT _____ LWT _____ GPM _____ PDPH _____
Heating Coil . . . . EWT _____ LWT _____ GPM _____ PDPH _____
Cooling Coil . . . . EWT _____ LWT _____ GPM _____ PDPH _____
Reheat Coil . . . . EWT _____ LWT _____ GPM _____ PDPH _____
For dual duct . . . . HDT _____ CDT _____
Discharge for multizone (zone temps· $^\circ$F) . . . .
z_1 _____ z_2 _____
z_3 _____ z_4 _____ z_5 _____ z_6 _____
z_7 _____ z_8 _____ z_9 _____ z_{10} _____

4. COMPONENT CONDITION (VISUAL INSPECTION)
Casing or Plenum . . . . Heavy Leaks _____ Medium Leaks _____ Nominal _____
Outside Air Louver . . . . Clean _____ Dirty _____ Clogged _____
Filters . . . . Clean _____ Dirty _____ Clogged _____
Filter Face Area _____ Ft2   Air Velocity Across Filter Face _____ GPM
Cooling Coil . . . . Clean _____ Dirty _____ Clogged _____
Heating Coil . . . . Clean _____ Dirty _____ Clogged _____

Figure 15-8. HVAC System Data
(Source: Certified Test & Balance Company, Inc., Chicago, Illinois)

Control Dampers
(Leakage in closed position) .... High ———— Normal ———— Low ————

Belts .... Tight ———— Loose ———— Worn ———— Good ————

5. AIR DISTRIBUTION

Is ductwork leaking? .... Heavy ———— Medium ———— Light ————

Is ductwork insulated/lined? .... Interior ———— Exterior ———— No ————

Is ductwork accessible to repair leaks? .... Yes ———— No ————

Does system have manual balancing dampers at zones or mains? .... Yes ———— No ————

Do supply outlets have dampers? .... Yes ———— No ————

Do return outlets have dampers? .... Yes ———— No ————

6. REMARKS:

_____

_____

_____

_____

_____

_____

_____

_____

_____

Figure 15-8. HVAC System Data (concluded)

PROJECT NO. _____

JOB NAME _____   DATE _____

SYSTEM NO. _____   LOCATION _____   O.A.T. _____

TESTED BY _____

EWT . . . . ACTUAL _____   DESIGN _____

LWT . . . . ACTUAL _____   DESIGN _____

COIL NO.	AREA SERVED	ACTUAL				DESIGN			
		CFM	EAT	LAT	AIR PD	CFM	EAT	LAT	AIR PD

REMARKS:

Figure 15-9. Reheat Coil Data
(Source: Certified Test & Balance Company, Inc., Chicago, Illinois)

DATE _____

PREPARED BY: _____

PROJECT NO. _____

JOB NAME:

AREA SERVED	GROSS SQ. FEET	SYSTEM SERVING	WEEKDAYS		SATURDAY		SUNDAYS/HOLIDAYS	
			NO. OF OCCU-PANTS	OCCUPANCY TIME A.M. P.M.	NO. OF OCCU-PANTS	OCCUPANCY TIME A.M. P.M.	NO. OF OCCU-PANTS	OCCUPANCY TIME A.M. P.M.

Figure 15-10. Building Occupancy Schedule
(Source: Certified Test & Balance Company, Inc., Chicago, Illinois)

JOB NAME _____

PREPARED BY _____

DATE _____

PROJECT NO. _____

SYSTEM NUMBER	DESIGN		ACTUAL		REMARKS:
	TOTAL CFM	OUTSIDE AIR	TOTAL CFM	OUTSIDE AIR	

NOTE: These readings should be obtained by traverse and O.S. setting kept on minimum position ONLY.

Figure 15-11. CFM Audit

(Source: Certified Test & Balance Company, Inc., Chicago, Illinois)

PROJECT NO. _____

JOB NAME _____ DATE _____

UNIT NO. _____ TYPE _____ (Steam-to-water/Water-to-water) _____

LOCATION _____ TESTED BY _____

	Actual	Design
Steam Pressure (PSIG) · · · · · · · ·	Actual _____	Design _____
Flow Rate (GPM) · · · · · · · ·	Actual _____	Design _____
Pressure Drop · · · · · · · ·	Actual _____	Design _____
EWT · · · · · · · ·	Actual _____	Design _____
LWT · · · · · · · ·	Actual _____	Design _____
Reset Control · · · · · · · ·	Automatic _____	Manual _____

REMARKS:

Figure 15-12. Hot Water Convertor Data
(Source: Certified Test & Balance Company, Inc., Chicago, Illinois)

PROJECT NO. _____

JOB NAME _____ DATE _____

SYSTEM NO. _____ MACHINE NO. _____ REFRIGERANT TYPE _____

LOCATION _____ TESTED BY _____

COOLER:

	Actual	Design
GPM Capacity · · · · · · · ·	Actual _____	Design _____
Pressure Drop · · · · · · · ·	Actual _____	Design _____
EWT · · · · · · · ·	Actual _____	Design _____
LWT · · · · · · · ·	Actual _____	Design _____

CONDENSER:

	Actual	Design
GPM Capacity · · · · · · · ·	Actual _____	Design _____
Pressure Drop · · · · · · · ·	Actual _____	Design _____
EWT · · · · · · · ·	Actual _____	Design _____
LWT · · · · · · · ·	Actual _____	Design _____

REMARKS:

Figure 15-13. Absorption Refrigeration Machine Data
(Source: Certified Test & Balance Company, Inc., Chicago, Illinois)

Surveyed by: _____

Survey Date: _____

1. **GENERAL INFORMATION**

IDENTITY:

OPERATION _____

Address _____

_____

Type(s) of occupancy _____

_____

Name of person in charge of energy _____

_____

PHYSICAL DATA:

Building orientation _____

No. of floors _____

Floor area, gross, square feet _____

Net air conditioned square feet _____

Construction type:

Walls (masonry, curtain, frame, etc.)

N _____ S _____ E _____ W _____

Figure 15-14. Building Information

Roof:

Type: Flat _____   Color: Light _____

     Pitched _____   Dark _____

Glazing:

Exposure	*Type	%Glass/Exterior wall area
N	_____	_____
S	_____	_____
E	_____	_____
W	_____	_____

*Type: Single, double, insulating, reflective, etc.

Glass shading employed outside (check one):

Fins _____ Overhead _____ None _____ Other _____

Glass shading employed inside (check one):

Shades _____ Blinds _____ Drapes, open mesh _____ Drapes opaque _____ None _____ Other _____

SKETCH OF BUILDING SHOWING PRINCIPLE DIMENSIONS:

BUILDING TYPE:

All electric _____

Gas total energy _____

Oil total energy _____

Other _____

BUILDING OCCUPANCY AND USE:

Weekdays: Occupied by:* _____ people from _____ to _____ (hours)

_____

_____

_____

_____

_____

Saturdays: _____

Sundays, holidays _____

Hours air conditioned: Weekdays from_____ to_____; Saturdays_____ to_____ Sundays, holidays from_____ to_____

*(Account for 24 hours a day. If unoccupied, put in zero)

2.   **ENVIRONMENTAL CONDITIONS**

OUTDOOR CONDITIONS

Winter: Day_____°F. dB_____ mph wind   Night_____°F. dB_____mph wind

Summer: Day_____°F. dB_____ mph wind   Night_____°F. dB_____mph wind

MAINTAINED INDOOR CONDITIONS:

Winter: Day_____°F. dB_____ %rh   Night_____°F. dB_____%rh

Summer: Day_____°F. dB_____ %rh   Night_____°F. dB_____%rh

Figure 15-14. Building Information (con't)

3.   **SYSTEMS AND EQUIPMENT DATA**

HVAC SYSTEMS:

Air handling systems (check as appropriate):

Perimeter system designation:

Single zone _____ Multizone _____

Fan coil _____ Induction _____

Variable air volume _____ Dual duct _____

Terminal reheat _____ Self-contained _____

Heat pump _____

Interior system designation:

Fan coil _____ Variable air volume _____

Single zone _____ Other (describe) _____

Principle of operation:

Heating-cooling-off _____

Air volume variation _____

Air mixing control _____

Temperature variation _____

Interior:

Heating-cooling-off _____

Air volume variation _____

Temperature variation _____

4.   AIR HANDLING UNIT – SUPPLY, RETURN, EXHAUST

System Description_____

Horsepower_____ OSA Dampers - Yes ☐   No ☐   M.A. Setting_____ °F

Location_____ Area Served_____

Terminal Units: Quantity_____ Type_____

Operations (Start-Stop)   Start Time   Stop Time

Monday thru Friday   _____   _____

Saturday   _____   _____

Sunday   _____   _____

Holiday   _____

Method of Start-Stop   Time Clock ☐   Manual ☐   Other ☐

Figure 15-14. Building Information (con't)

5. AIR HANDLING UNIT – SUPPLY, RETURN, EXHAUST

System Description _____

_____

_____

Horsepower _____ OSA Dampers - Yes ☐   No ☐   M.A. Setting _____ °F

Location _____ Area Served _____

Terminal Units: Quantity _____ Type _____

Operations (Start-Stop)   Start Time   Stop Time

Monday thru Friday   _____   _____

Saturday   _____   _____

Sunday   _____   _____

Holiday

Method of Start-Stop   Time Clock ☐   Manual ☐   Other ☐

6.   COOLING PLANT

Chillers: Number _____ Total Tonnage/KW _____

Chilled Water Pumps _____ Total HP _____

Condensed Water Pumps _____ Total HP _____

Cooling Tower Fan(s) _____ Total HP _____

Chilled Water Supply Temp., Setpoint _____ °F

Operations (Start-Stop)   Start Time   Stop Time

Monday thru Friday   _____   _____

Saturday   _____   _____

Sunday   _____   _____

Holiday   _____   _____

Method of Start-Stop   Time Clock ☐   Manual ☐   Other ☐

Figure 15-14. Building Information (con't)

Months Operation per Year _____

Remarks _____

7.   BOILER PLANT

Boiler No. _____ Size _____ Type _____

Fuel Used _____

Hot Water Supply Setpoint _____ F   Steam Pressure Setpoint _____ psi

Number of Pumps _____ Total HP _____

Remarks _____

8.   ROOFTOP/UNITARY SYSTEMS

Manufacture and Model _____

Quantity _____ Location _____

Cooling Capacity _____ Tons Total _____

Heating Capacity _____ Btu Output _____ Btu Input (Gas/Oil) _____

Electric ☐   Gas ☐   Steam/HW ☐

Single Zone Units _____ Multizone Units _____ Number of Zones _____

O.A. Damper Control _____

Fans: CFM   HP

Supply   _____   _____

Return   _____   _____

Exhaust   _____   _____

Operations   Start Time   Stop Time

Monday-Friday   _____   _____

Saturday   _____   _____

Sunday   _____   _____

Holiday   _____   _____

Figure 15-14. Building Information (con't)

Method of Start-Stop   Time Clock ☐   Manual ☐   Other ☐

9. **EXHAUST, AIR, MAKEUP AIR SYSTEMS**

Designation	Location	Area Served	CFM	HP
_____	_____	_____	_____	_____
_____	_____	_____	_____	_____
_____	_____	_____	_____	_____

TOTAL _____

Operating Schedule _____

All fans (supply, return and exhaust):

Location	Horsepower	Type	Method of Operation
_____	_____	_____	_____
_____	_____	_____	_____
_____	_____	_____	_____

Source of heating energy:

Hot water_____ Steam_____ Electric resistance_____ Other_____

Heating plant:

Boiler No._____ Rating_____ MBH

_____

_____

Boiler type:

Firetube_____ Watertube_____ Elec. resist._____ Electrode_____ Other_____

Fuel used_____ Standby_____

Hot water supply_____ °F, Return_____ °F

Steam pressure_____ psi

Pumps No._____ Total HP_____

Room heating units:

Type: Baseboard_____ Convectors_____ Fin tube_____

Ceiling or wall panels_____ Unit heaters_____ Other_____

Cooling plant:

Chillers: No._____ Total capacity (tons)_____

Type: Centrifugal_____ Reciprocating_____ Absorption_____

Figure 15-14. Building Information (con't)

Capacity controlled by: _____

Chiller operation: Starting controls _____

Stopping controls _____

Chilled water temp. supply_____°F, return_____°F

Condenser water temp._____ in °F_____ out °F

Heat dissipation device:

Evaporative condenser _____

Air cooled condenser _____

Cooling tower _____

Condenser/cooling tower fan HP_____

Heat recovery device: Double bundle condenser_____ Other_____

Chilled water pumps_____ Total HP_____

Condenser water pumps_____ Total HP_____

Self-contained units:

Type: Thru-the-wall-air conditioner_____ Other_____

No. of units_____ Basic module served_____

Capacity (tons)_____

10. **ENERGY CONSERVATION DEVICES:**

Type:

Condenser water used for heating_____

Demand limiters_____

Energy storage_____

Heat recovery wheels_____

Enthalpy control of supply-return-exhaust damper_____

Recuperators_____

Others_____

LIGHTING:

Interior lighting type:_____

Watts/ft^2: Hallway/corridor_____

Work stations_____

Circulation areas within work space_____

On-off from breaker panel_____ Wall switches_____

Control switching_____

Exterior Lighting: Type_____ Total KW_____

DOMESTIC HOT WATER HEATING:

Size_____ Rated input_____ Water Temp._____ °F

Energy Source: Gas_____, Oil_____, Electric_____, Other_____

Figure 15-14. Building Information (con't)

OTHER EQUIPMENT (Kitchen, etc.):

Equip. Description	Quantity	Size/Capacity in BTU, KW, HP, etc.

11. OPERATING SCHEDULE:

OPERATION (Start-stop)

Equipment description	Weekdays	Saturday	Sunday	Holiday
Refrigeration cycle mach.				
Fans – supply				
Fans – return/exhaust				
Fans – exhaust only				
HVAC auxilliary equip.				
Lighting – interior				
– exterior				
Fan kitchen exhaust				
Elevators				
Escalators				

Domestic hot water ht. _____ _____

Other (describe: _____) _____ _____

12. **LIGHTING**

1. Interior Lighting Type _____

   Watts/Ft.^{2} Offices _____ Other _____

   Total Install KW _____ Foot Candles _____

   On-Off from Breaker Panel? _____

   Wall Switch? _____ Control Switching? _____

   Operating Schedule _____

2. **Exterior Lighting Type** _____

   Total KW _____

   Operating Schedule _____

3. Remarks _____

   _____

Figure 15-14. Building Information (con't)

13.   **UTILITIES**

Electric Utility _____

        Rate Schedule _____ Effective _____

        Name of Rep _____ Phone _____

Gas Utility _____

        Rate Schedule _____ Effective _____

        Name of Rep _____ Phone _____

Water Utility _____

        Rate Schedule _____ Effective _____

        Name of Rep _____ Phone _____

EMERGENCY GENERATORS

Number _____ Size _____ KW

How Started: Manual ☐   Auto Switchover ☐

Equipment/Systems Operated: _____

_____

_____

CHECK LIST

	Due Date	Date Complete	By
1. HVAC Survey	_____	_____	_____
2. Lighting & Misc. Survey	_____	_____	_____
3. Utility Bill Analysis	_____	_____	_____
4. Recommendation	_____	_____	_____

Date _____

Figure 15-14. Building Information (concluded)

SUMMARY SHEET

GENERAL INFORMATION
Building/Plant/Business Center: _____
Address: _____
City, State: _____
Building Supervisor: _____
Building Use: _____

TOTALS BY BUILDING LOCATION:

Building Location	Total Area Building Location	Allowance Sq Ft	Total Watt Allowance	Total Connected Load
1		3.0		
2		1.0		
3		0.5		
Interior Total				
		Allowance Ft		
4		5.0		
5		0.5		
Exterior Total				

BUILDING LOCATION DESIGNATIONS:
1 = Office Space/Personnel
2 = Rest, Lunch, Shipping/Warehouse
3 = Malls, Lobby
4 = Building Perimeter, Facade, Canopy
5 = Parking

DETAIL SHEET

GENERAL INFORMATION
Building/Plant/Business Center: _____
Address: _____
BUILDING LOCATION USE: _____

Room Name	Area Sq Ft	Lamp or Fixture Type	Quantity	Watts Unit	Total Connected Load (Watts)
TOTAL					

Figure 15-15. Lighting Audit

TYPICAL LIGHTING FIXTURE WATTAGE

I. FLUORESCENT

Lamp Description	Lamps per / Lamp Fixture / Type	Fixture Wattage*
4 Ft 40 Watt Rapid Start	1-F40T12 2-F40T12 3-F40T12 4-F40T12	50 92 142 184
8 Ft Slimline Instant Start	1-F96T12 2-F96T12 3-F96T12 4-F96T12	100 170 270 340
8 Ft High Output	1-F96T12/HO 2-F96T12/HO 3-F96T12/HO 4-F96T12/HO	140 252 392 504
8 Ft 1500 ma Power Grove, SHO or VHO	1-F96PG17 1-F96T12/SHO or VHO 2-F96PG17 2-F96T12/SHO or VHO 3-F96PG17 3-F96T12/SHO or VHO 4-F96PG17 4-596T12/SHO or VHO	230 450 680 900

II. HIGH INTENSITY DISCHARGE

Lamp Type	Lamp Designation	Watts	Fixture Wattage*
Mercury	MV 100 MV 175 MV 250 MV 400 MV 1000	100 175 250 400 1000	118 200 285 450 1075
Metal Halide	MH 175 MH 250 MH 400 MH 1000	175 250 400 1000	210 292 455 1070
High Pressure Sodium	HPS 70 HPS 100 HPS 150 HPS 250 HPS 400 HPS 1000	70 100 150 250 400 1000	88 130 188 300 465

* Includes Lamp and Ballast Wattage.

Figure 15-15. Lighting Audit (con't)

To plan more adequate lighting, refer to the table below which illustrates the energy used by various types of lighting.

LUMEN/WATT TYPICAL LIGHT SOURCES

Source	Initial Lumens Watt
Low pressure sodium (35 to 180 watts)	133 to 183
High pressure sodium (70 watts to 1,000 watts)	83 to 140
Metal halide (175 watts to 1,000 watts)	80 to 125
Fluorescent (30 watts to 215 watts)	74 to 84
Mercury (100 watts to 1,000 watts)	42 to 62
Incandescent (100 watts to 1,500 watts)	17 to 23

RECOMMENDED REFLECTANCE VALUES

Surfaces	Reflectance (Percent)
Ceiling	80-90%
Walls	40-60%
Desks and Bench Tops, Machines and Equipment	25-45%
Floors . . . . .Not less than 20%	

Figure 15-15. Lighting Audit (concluded)

MFG'R.	LIGHT # FIXTURE	LOCATION	NO.	WATTS PER FIXTURE	LUMENS	Hrs. Operated Per Day	Days Operated Per Week	KWH Per Per Week	COMMENTS

OPERATION _____ LOCATION _____ DATE _____

Figure 15-16. Energy Survey – Lights

OPERATION _____

LOCATION _____

DATE _____

MFG'R.	EQUIPMENT ITEM	LOCATION	ELECTRICAL EQUIPMENT RATED INPUT						COMMENTS
			1 AMPS	2 VOLTS	3 KW* 1 x 2 1000	4 HRS. OPERATED PER DAY	5 DAYS OPERATED PER WEEK	6 KWH PER WEEK 3 x 4 x 5	

1. Fuel or power requirement is usually listed on equipment name plate.
2. To find Btu's per hour for
   Electricity— multiply amps x volts x 3.413 or watts x 3.413
   Natural Gas— multiply cubic feet per hour x 1,000
   #2 Fuel Oil— multiply gallons per hour x 140,000
   Steam— multiply pounds per hour x 1,000
3. To find cost per hour, multiply the cost per Btu for that unit's fuel (from Worksheet   x the Btu's per hour by that unit.

*To find Btu equivalents (expressed in millions) for electricity, multiply kilowatt hours x .003414.
For natural gas, multiply cubic feet x .001.
For #2 fuel oil, multiply gallons x .14.
For purchased steam, multiply pounds x .001.
To find the cost per million Btu's in each category, divide the total cost in that category by the total Btu's, expressed in millions, for that category.

Figure 15-17. Energy Survey — Electrical Equipment

METER NO. _____

LOCATION _____

STATE _____

MONTHLY ENERGY USE _____

OPERATION _____

MONTH	CONSUMPTION KWH	DEMAND KW	RATE	FUEL ADJ. RATE/KWH	COST	65° HEATING DAYS	65° COOLING DAYS	KWH INCREASE OVER PAST YEAR	COST INCREASE OVER PAST YEAR	COST INCREASE FUEL ADJ. RATE OVER/UNDER PAST YEAR
JAN										
FEB										
MAR.										
APR.										
MAY										
JUNE										
JULY										
AUG.										
SEPT.										
OCT.										
NOV.										
DEC.										
TOTALS										

	JAN.	FEB.	MAR.	APR.	MAY	JUNE	JULY	AUG.	SEPT.	OCT.	NOV.	DEC.	COMMENTS
% INCREASE COST													
% INCREASE KWH													
% INCREASE FUEL ADJ. RATE													
BASIC RATE INCREASE													

Figure 15-18. Electrical Worksheet

OPERATION _____   LOCATION _____   DATE _____

MFG'R.	EQUIPMENT ITEM	LOCATION	GAS EQUIPMENT RATED INPUT					COMMENTS
			7 BTU PER HOUR	8 CU. FT. PER HR. 7/1000	9 HOURS OPERATED PER DAY	10 DAYS OPERATED PER WEEK	11 CU. FT. PER WEEK 8 x 9 x 10	

1. Fuel or power requirement is usually listed on equipment name plate.
2. To find Btu's per hour for:
   Electricity— multiply amps x volts x 3.413 or watts x 3.413
   Natural Gas— multiply cubic feet per hour x 1,000
   #2 Fuel Oil— multiply gallons per hour x 140,000
   Steam— multiply pounds per hour x 1,000
3. To find cost per hour, multiply the cost per Btu for that unit's fuel (from Worksheet   x the Btu's per hour by that unit.

*To find Btu equivalents (expressed in millions) for electricity, multiply kilowatt hours x .003414.
For natural gas, multiply cubic feet x .001.
For #2 fuel oil, multiply gallons x .14.
For purchased steam, multiply pounds x .001.
To find the cost per million Btu's in each category, divide the total cost in that category by the total Btu's, expressed in millions, for that category.

Figure 15-19. Energy Survey — Gas Equipment

METER NO. _____

LOCATION _____

STATE _____

OPERATION _____

MONTHLY ENERGY USE _____

MONTH	CONSUMPTION CCF	RATE	FUEL ADJ. RATE/CCF	COST	65° HEATING DAYS	65° COOLING DAYS	CCF INCREASE OVER PAST YEAR	COST INCREASE OVER PAST YEAR	COST INCREASE FUEL ADJ. RATE OVER/UNDER PAST YEAR
JAN.									
FEB.									
MAR.									
APR.									
MAY									
JUNE									
JULY									
AUG.									
SEPT.									
OCT.									
NOV.									
DEC.									
TOTALS									

	JAN.	FEB.	MAR.	APR.	MAY	JUNE	JULY	AUG.	SEPT.	OCT.	NOV.	DEC.	COMMENTS
% INCREASE COST													
% INCREASE CCF													
% INCREASE FUEL ADJ. RATE													
BASIC RATE INCREASE													

Figure 15-20. Gas Worksheet

MONTHLY ENERGY USE

METER NO._____
LOCATION_____
STATE_____
OPERATION_____

MONTH	CONSUMPTION GALLONS	RATE	FUEL ADJ. RATE/GAL.	COST	65° HEATING DAYS	65° COOLING DAYS	GAL. INCREASE OVER PAST YEAR	COST INCREASE OVER PAST YEAR	COST INCREASE FUEL ADJ. RATE OVER/UNDER PAST YEAR
JAN.									
FEB.									
MAR.									
APR.									
MAY									
JUNE									
JULY									
AUG.									
SEPT.									
OCT.									
NOV.									
DEC.									
TOTALS									

	JAN.	FEB.	MAR.	APR.	MAY	JUNE	JULY	AUG.	SEPT.	OCT.	NOV.	DEC.
% INCREASE COST												
% INCREASE GALLONS												
% INCREASE FUEL ADJ. RATE												
BASIC RATE INCREASE												

COMMENTS

Figure 15-21. Fuel Oil Worksheet

METER NO. _____
LOCATION _____
STATE _____

MONTHLY ENERGY USE _____

OPERATION _____

MONTH	CONSUMPTION LBS.	RATE	FUEL ADJ. RATE/LBS.	COST	65° HEATING DAYS	65° COOLING DAYS	LBS. INCREASE OVER PAST YEAR	COST INCREASE OVER PAST YEAR	COST INCREASE FUEL ADJ. RATE OVER/UNDER PAST YEAR
JAN.									
FEB.									
MAR.									
APR.									
MAY									
JUNE									
JULY									
AUG.									
SEPT.									
OCT.									
NOV.									
DEC.									
TOTALS									

	JAN.	FEB.	MAR.	APR.	MAY	JUNE	JULY	AUG.	SEPT.	OCT.	NOV.	DEC.
% INCREASE COST												
% INCREASE LBS.												
% INCREASE FUEL ADJ. RATE												
BASIC RATE INCREASE												

COMMENTS

Figure 15-22. Steam Worksheet

METER NO._____

LOCATION_____

STATE_____

MONTHLY ENERGY USE _____

_OPERATION_____

MONTH	CONSUMPTION GALLONS	RATE	ADJ. RATE/GAL.	COST	65° HEATING DAYS	65° COOLING DAYS	GAL. INCREASE OVER PAST YEAR	COST INCREASE OVER PAST YEAR	COST INCREASE ADJ. RATE OVER/UNDER PAST YEAR
JAN.									
FEB.									
MAR.									
APR.									
MAY									
JUNE									
JULY									
AUG.									
SEPT.									
OCT.									
NOV.									
DEC.									
TOTALS									

	JAN.	FEB.	MAR.	APR.	MAY	JUNE	JULY	AUG.	SEPT.	OCT.	NOV.	DEC.	COMMENTS
% INCREASE COST													
% INCREASE GALLONS													
% INCREASE ADJ. RATE													
BASIC RATE INCREASE													

Figure 15-23. Water Worksheet

Building or location _____

Type (check below)

    Steam boiler _____   Hot water generator _____

    Hot air furnace _____   Other _____

Fuel source: Natural gas _____   #1 Oil _____

            Butane _____   #2 Oil _____

            Propane _____   #6 Oil _____

                                    Other _____

Rated pressure of boiler or generator _____

Measured water or steam system, pressure drop _____ psig

Pump motor: Voltage _____   Amperage _____

            Manufacturer _____   Phases _____

Minimum pressure drop, assuming no corrosion or fouling _____ psig

Nameplate or rated output _____ Btu/hr; Hp

Design heat loss of system _____ Btu/hr

Measured draft pressure ±_____ in. H_2O

Location of measurement:

_____ Over-the-fire _____ Breaching

Type of draft: _____ Forced _____ Induced

Acceptable draft pressure ± _____ in. H_2O
   (Refer to table at end of form)

Measured smoke density reading
   (For oil burners only)

Measured CO_2 concentration _____ %

Acceptable CO_2 range _____ %
   (Refer to table at end of form)

Measured stack temperature _____ °F

Measured make-up air (or boiler room air) temperature _____ °F

Net stack temperature = _____ °F

Acceptable net stack temperature _____ °F

Measured boiler efficiency _____ %

RECOMMENDED DRAFT PRESSURES (IN. H_2O)
FOR COMBINATION SYSTEMS

	Location	
Gas or Oil Burners	Over the Fire	Boiler Breaching
Natural or induced draft	−0.02 to −0.05	−0.07 to −0.10
Forced draft	0.70 to 0.10	0.02 to 0.05

Figure 15-24. Combustion System Data
(Source: Manual of Procedures for Authorized Class A Energy Auditors in Iowa)

**APPROXIMATE STOICHIOMETRIC AND RECOMMENDED CO$_2$
CONCENTRATIONS FOR VARIOUS FUELS**

		% CO$_2$	
Gases	Stoichiometric		Recommended Value
Natural	12		7-10
Propane or Butane	14		8.5-11.5
Fuel Oils			
No. 1 or No. 2	15		9-12
No. 6	16.5		10-14

Figure 15-24. Combustion System Data (concluded)

Table 15-2. List of Conversion Factors

1 U.S. barrel	= 42 U.S. gallons
1 atmosphere	= 14.7 pounds per square inch absolute (psia)
1 atmosphere	= 760 mm (29.92 in) mercury with density of 13.6 grams per cubic centimeter
1 pound per square inch	= 2.04 inches head of mercury
	= 2.31 feet head of water
1 inch head of water	= 5.20 pounds per square foot
1 foot head of water	= 0.433 pound per square inch
1 British thermal unit (Btu)	= heat required to raise the temperature of 1 pound of water by 1°F
1 therm	= 100,000 Btu
1 kilowatt (Kw)	= 1.341 horsepower (hp)
1 kilowatt-hour (Kwh)	= 1.34 horsepower-hour
1 horsepower (hp)	= 0.746 kilowatt (Kw)
1 horsepower-hour	= 0.746 kilowatt hour (Kwh)
1 horsepower-hour	= 2545 Btu
1 kilowatt-hour (Kwh)	= 3412 Btu
To generate 1 kilowatt-hour (Kwh) requires 10,000 Btu of fuel burned by average utility	
1 ton of refrigeration	= 12,000 Btu per hr
1 ton of refrigeration requires about 1 Kw (or 1.341 hp) in commercial air conditioning	
1 standard cubic foot is at standard conditions of 60°F and 14.7 psia.	
1 degree day	= 65°F minus mean temperature of the day, °F
1 year	= 8760 hours
1 year	= 365 days
1 MBtu	= 1 million Btu
1 Kw	= 1000 watts
1 trillion barrels	= 1 × 10^{12} barrels
1 KSCF	= 1000 standard cubic feet

Note: In these conversions, inches and feet of water are measured at 62°F (16.7°C), and inches and millimeters of mercury at 32°F (0°C).

16

Computer Software for Energy Audits

James P. Waltz, President
Energy Resource Associates,
Livermore, CA

INTRODUCTION

This chapter intends to visit the subject of computer software as it applies to or contributes to the performance of energy audits in commercial buildings. In doing so, the chapter will attempt to overview the various uses to which software could be put to use in an energy audit and then focus in on the use which we believe is perhaps the most important (building simulation), offering in the process both some general philosophies regarding building simulation and some very specific guidance and suggestions on how to perform building simulation in a practical and highly accurate fashion.

SOFTWARE..., THE POSSIBILITIES

It seems that personal computers and their software have invaded our lives. Everywhere we turn there's a new program or capability that we just can't live without (or at least the competition has convinced our mutual customers that this is so!). Well, the energy auditing business is no different than the rest of the world. In fact, given the technical nature of the business of energy auditing, it's perhaps not surprising that computers have such a significant impact on the process, in the following possible ways.

Energy Accounting

While discussed elsewhere in this book, energy accounting is a particular activity that lends itself to computerization. A range of "off the shelf" software is available here, as is advertised in the energy trade publications, and also as offered by such sources as Elite Software and Northridge Engineering Software, who also advertise in the trade publications. In addition, both ASHRAE and Energy Engineering regularly publish catalogs of energy-related software, which list energy accounting programs.

Whether a home-built spreadsheet or a stand-alone program, energy accounting software can provide a number of functions, including:

- tabulating large quantities of energy use data

- pro-rating the data so as to provide calendar-month consumption figures (as opposed to varying-length billing periods)

- calculating energy use (Btu/sf/yr) and energy cost (S/sf/yr) indexes, which are useful for comparing buildings against each other and other established "norms"

- showing recent trends in energy use (is energy use going up or down?) accounting for savings achieved by an energy retrofit program, including documenting and adjusting for the effects of weather and other independent variables

- calculation of average unit costs and showing trends in same

Our belief is that no energy conservation project should begin or end without passing through the energy accounting process.

Survey Data Reduction

While this use of software is somewhat minor and perhaps obscure, it is nonetheless very important. A technically rigorous energy audit should include a significant amount of measurements of the actual operating parameters of the installed equipment. Since nearly all energy-using systems in buildings use electricity, instantaneous electrical measurements of the power draw of lighting panels, receptacle panels, motors, chillers, computer rooms and other process loads can amount to a prodigious quantity of data. Our experience has shown that calculation of the actual kw of these measurements (from measured volts, amps and power factor—we recommend against kw meters) utilizing the computer instead of manual calculations has numerous beneficial effects, including:

•   minimizes calculational errors

•   provides reliable and neatly organized data for use in analysis and post-retrofit troubleshooting

•   shows the client that the analysis has been accomplished in a well-organized and professional fashion, especially if the printouts are included in the final report

Either home-grown or professionally-developed/automated spreadsheets can be effective in this activity.

Cost Estimating

Only three issues must really be addressed when considering energy retrofit; technical feasibility, cost of installation and probable energy savings. While many practitioners perform their cost estimating by hand, the well-known estimating guides (Means for example) offer complete software programs to perform a wide range of estimating tasks. Those involved in large quantities of energy retrofit estimating would be well to consider utilizing an automated method of estimating.

Computer Assisted Design and Drafting (CADD)

Many people would be slow to recognize that CADD has a significant contribution to offer to the energy auditing process. A large recent project brought home the potential for utilizing CADD in the energy auditing process in the following ways:

- it is always helpful to have reduced-size floor plans to assist in finding the way around a facility (especially a large and complex facility) having the building floor plans on CADD allows printing out whole or partial floor plans to whatever size is convenient for use during the field survey

- lighting surveys, especially when reflected ceiling plans are not available, can be greatly assisted by printing out small-size partial floor plans for recording fixture types and counts during the field survey

- during development of retrofit measures, some equipment (chiller, cogeneration units, etc.) require significant "real estate" for their installation—printing out a portion of the building floor plan can greatly assist in planning the equipment installation and documenting it in the project report/files

- building simulation usually requires "zoning" the building for eventual building dimension take-off and data entry into the simulation program using CADD-produced, to-scale floor plans can greatly facilitate this process

Again, though it is not an "obvious" energy audit tool, there are many contributions that may be made by CADD software.

Estimating Energy Savings

While some practitioners still perform savings calculations the old-fashioned way, by hand, the majority of savings calculations are automated to at least a minimal extent at this time. Such calculations, when not part of a building simulation program (discussed below), are based on a large number of simplifying assumptions and utilize fairly simplistic formulae. In some cases these "stand-alone" calculations are highly automated and are based on extensive weather files and other elaborate data sources. Such calculations depend upon pre-established spreadsheet (Lotus, Excel, etc.) or math program (e.g.,

Mathcad) files, or are implemented through stand-alone programs. The advantages of such automated calculations are standardization, speed, ease of use and generally good documentation. However, they suffer from a lack of context. That is, because they stand alone, these calculations may easily overstate potential savings because they exist "outside" of a building energy balance—which would otherwise constrain the calculation to the actual energy consumed by a given end-use.

Building Energy Simulation

Originating around the mid-1970's, computers simulation of building energy use was developed as a tool for analyzing buildings during their initial design to develop load estimates and optimal combinations of building features. The technique has possibly found even better use in analyzing existing buildings for energy conservation retrofit. By simulating retrofit options on the computer, reliable estimates of potential energy savings may be achieved, assuming that the initial modeling and subsequent modeling of retrofit measures has been correctly performed. In fact, the author has used a wide range of computer simulation tools over the past decade to prepare savings estimates for a large number of comprehensive energy retrofits in both large (1.8 million square feet) and small (25,000 square feet) buildings with great success.

In at least one case, a major energy services company declined to accept the conservative savings figures generated by computer modeling of the building, implemented a project based on their own optimistic estimates of savings, and ended up reimbursing their client for more than $100,000 annually for the project's "shortfall" in savings (from their optimistic estimates). Their optimistic estimates increased total estimated savings by more than $200,000 per year— fortunately (for them), the energy services company did not guarantee 100% of their estimated savings.

The importance of accurate modeling of existing buildings is clearly critical to the business of energy services, demand-side management, or any form of energy retrofit. This is even more important given the growing employment of demand-side management as a supply strategy by utility companies nationwide.

BUILDING ENERGY SIMULATION—
WHAT AND WHY

In the building simulation phase of an energy audit, all the data gathered during the field survey is converted into a form acceptable to the computer. The building's architectural and functional use characteristics are described, including orientation and thermal properties of the structure ("U values and shading coefficients, for example). Thermal zones or spaces in the facility having similar external and internal characteristics are designated, and aggregate schedules and thermal loads for people, lights and equipment developed. The space heating and cooling system characteristics are described, including the mode of operation and control (e.g., multi-zone versus terminal reheat, supply air temperatures, economizers, etc.) air flow rates, operating schedules and assignment of building thermal zones to individual systems. Central equipment (boilers, chillers, etc.) that serve space-conditioning systems are described, including the equipment type (e.g., water tube versus fire tube boilers), efficiencies, method of control, operating schedules, and flow rates.

Once the simulation model is assembled it is run on the computer (in some cases taking as much as 6 hours for a single complex building, even when using a fast P.C.!) and is then verified by comparing it to the actual energy use of the building, as recorded by the utility company. In addition, the model is examined in detail to confirm that individual components are faithful to known physical realities of the building. For example, if chiller electrical use was measured during the survey, this actual measured energy use of the chiller would be contrasted to the model to confirm that cooling loads were being faithfully simulated in the model. Only when the model agrees with reality in terms of total annual energy use (less than 10% variation), seasonal patterns of energy use and individual component energy use, should the model be considered complete.

The model then, serves two purposes. First of all, it serves as an energy balance in that it accounts for all sources and uses of energy in the building. As such, the model cannot be faithful unless the investigators' knowledge of the building and how it operates agrees with reality, and creating the model "tests" the investigators' knowledge and forces that knowledge to be added to or corrected until it agrees with and encompasses the truth (as revealed in the utility company invoices) about the building being modeled. Secondly, the model, once verified, serves as a "test bed" for examining the effectiveness of energy retrofit modifications being considered for a building. Literally, these modifications can be tested, before being built, by simulating them on the model.

BUILDING ENERGY SIMULATION— WHAT THEY DON'T TELL YOU

The business of building energy simulation has been around for about two decades now, and has developed an image and reputation all its own. However, the image is often quite distant from the reality, particularly with respect to energy simulation of existing buildings.

The History of Building Simulation

In the beginning, building simulation was oriented towards new building design (remember two decades ago when new construction was virtually "the industry"?). All the programs had to do was perform a good load calculation and simulate a fairly narrow range of HVAC system types that were expected to operate the way they were designed to operate. Pretty simple, right?

This was appropriate at the time because the time to get such things as the building envelope correct is when it hasn't been constructed yet. As new construction was the "frontier" in this business at that time, efforts spent on developing building energy codes and certification of computer programs for use in demonstrating code-compliance were focused on new construction. Furthermore, since building codes and new frontiers are a specialty of government agencies and the R&D side of the industry, the people doing the lion's share of computer program development and program certification had little or no experience in the business of designing and building real buildings (for example, one firm acknowledged as experts in the field, cannot show any actual construction project experience at all among it's entire resume of professionals!). The result, was computer programs, "certified" or not, that were poorly suited for analysis of existing buildings—buildings where control systems function not as intended, where building occupancy is far from "neat and tidy," where HVAC systems frequently have little resemblance to the "models" offered in the simulation programs (how about a single air handling system combining variable volume, multi-zone and high-pressure induction?), and where building equipment operating practices are anything but consistent. Such factors are far more important in determining the energy use of an existing building than the precision with which we are able to model the insulation value of an exterior wall.

Evidence of this highly theoretical nature of the programs and the "profession" include the sprinkling of articles in the literature regarding "shoot-offs" and other "blind" comparisons of one building simulation program against another. These are all meaningless relative to existing buildings, as we are not "blind" when working with existing buildings—we can observe and develop a body of reasonably accurate knowledge regarding the real connected loads, real occupancy patterns, real system and equipment operating characteristics, etc., and use this body of knowledge to build and calibrate accurate and (most importantly) useful models of existing buildings.

The only correct conclusion that can be drawn from all this is that there is not and can not be such a thing as a "perfect" building energy simulation computer program. There can only be a well trained, knowledgeable and experienced professional who employs the tools in his/her possession in a professional manner.

The State of the Art

Newcomers to the field (and "innocent" bystanders—frequently a practitioner's clients) are frequently awed by the sophistication and complexity of the building simulation process and the computer programs employed in the process. However, the naked truth is that a very large number of practitioners of building simulation are woefully ignorant of its correct use and are very often guilty of perpetuating the "garbage-in-garbage-out" syndrome. One reason for this is that, unless modeling critique and calibration procedures as discussed later in the chapter are followed, a perfectly "plausible" simulation can be prepared that is largely fallacious—often with the ignorance of the building energy simulation practitioner themselves. To make matters worse, few receivers of such services, i.e., clients, are in a position to question, let alone critique, such a work product. A principal underlying cause of this situation is that virtually all of the training that is available on the subject of building energy simulation is focused on the software programs themselves, not the process of building energy simulation. As we will have implied above and will see later below, the software itself is perhaps the least important part of the building energy simulation process (an assertion that, however true it may be, is sure to create consternation among the leading professionals in the building energy simulation "profession").

BUILDING
ENERGY SIMULATION—
HOW TO DO IT RIGHT

For a computer simulation of a building to be of value in evaluating energy retrofit opportunities, it must be accurate. To be "accurate," the model should account for essentially all of the sources and uses of energy in a building. Such a model would calculate a total energy consumption that is close to the building's actual annual energy use, say within 5%. Such a model would also reasonably accurately mirror the building's actual response to the changing seasons of the year and closely mimic the actual seasonal variations in energy used by the building. Finally, such a model would allocate energy use by function in a faithful fashion. This last virtue is particularly important if the model is to be used to evaluate the effectiveness of specific energy retrofit measures, for example, lighting controls, outside air economizers, etc.

Accuracy in computer simulation of buildings, in our experience, is founded in three basic areas:

1. an intimate understanding of the simulation tool being used, including its various idiosyncrasies and nuances;

2. an intimate understanding of the building being simulated, vis-a-vis its physical and operational characteristics—in essence, in existing buildings, the quality of the survey or "audit" determines the quality of the simulation;

3. careful analysis and critique of output data (just because it is carefully prepared and computer generated doesn't mean it is correct)—our comments elsewhere herein generally apply to "mainframe" programs, though they also apply to other simulation tools.

By utilizing the above techniques, we have found it possible to regularly model buildings within 5% of their actual annual energy use with a high degree of confidence in the simulation of each energy-using system and functional use of energy in the building. It should be noted that, in buildings where weather is a strong energy-use factor, modeling to less than 10% variance from the actual energy use may be of limited value as our ability to predict weather for a given future year may not even be that accurate.

Knowledge of the Tool

The first foundation of accurate building simulation is knowledge of the tool to be employed.

While the above statement may seem obvious, the computer simulation tools available to consulting engineers are very complex and have a "reality" of their own that cannot be ignored or violated if accurate models of buildings and energy retrofit measures are to be accomplished.

Very specific experience comes to mind in this regard having to do with assignment of lighting loads and quantities of outside air. It is fairly common to utilize return air troffer lighting fixtures to reduce the in-space load on supply air, thus allowing a lower supply air quantity and a raising of return air temperature that allows selection of a smaller cooling coil for the same cooling capacity. In one project during the design development stage, the engineer was modeling an office building that had a large amount of core space that served mostly as trafficways, secretarial space, and file/storage space. As a result, the principal cooling load was created by the lighting systems. Unfortunately, the design engineer assigned virtually all of the lighting load to the return air (which is not physically possible) and specified a minimum outside air quantity as a cfm/ft^2 figure. The result was that supply air was calculated by the program at something like 0.2 cfm/ft^2, the outside air was 0.1 cfm/sf^2 and the computer calculated a return air temperature of around 500 degrees. When half this return air was discarded, roughly half the cooling load went with it, for an amazing "savings" in energy use. Upon detailed examination of the computer output, we were able to point out the fallacy of the simulation and got the project back on track. The experience did make the point, however, that a lack of detailed understanding and familiarity with the calculational methodology of the simulation program can easily lead the modeler astray!

Another example is the capability or lack of capability to handle desired simulations by the program. Before variable-flow chilled-water pumping was commonly employed, few programs had the ability to simulate such a system. In order to do so, a series of "dummy" chillers were described to the program in such a manner that the program selected each in turn as loads increased. Associated with each of these "dummy" chillers was a constant speed and power pump. The effect of each chiller/pump combination was to sequentially simulate the overall pump power curve that would be produced by a single variable speed pump. For accurate estimation of savings, only the energy consumed by the pumping systems was compared from one run to the next, thereby eliminating unwanted secondary impacts such as changes in chiller efficiency.

To be knowledgeable about a simulation program, the user must understand how the input data is understood and utilized by the program, the calculations/algorithms employed by the program, the flow of input and calculated values through the program, and the precise effect various program "controls" exert on the calculations performed by the program. The bottom line here is that an inferior simulation tool in the hands of an engineer well versed in its features and capabilities is superior to the best simulation tool in the hands of an engineer unfamiliar with it.

Knowledge of the Building

Perhaps the single most important factor in developing accurate computer models of existing buildings is developing an intimate knowledge of the physical and operational characteristics of the building to be modeled.

Envelope and Weather Versus Operators and Controls

While many practitioners of computer simulation of buildings work toward more detailed time-related simulation of weather its effect on building structures, those who are well acquainted with the practical aspects of building operation know that the effect of operating engineers and temperature control systems are manyfold more dominant in affecting a building's energy use. Perhaps one or two anecdotes would be illustrative of this point.

In one study of a major high-rise office building in San Francisco, it was observed late one evening that the watch engineer was "fiddling" with the central temperature control panel. Immediately thereafter, the indicating instruments on the panel all began to change their values rapidly. Gently interrogating the watch engineer, it was learned that the "fiddling" was to put the outside air economizer control for the entire building back on "automatic." Further investigation revealed that it was this engineer's nightly practice to override these controls to place all operating HVAC systems (a few terminal reheat systems serving the entire core of the building) on 100% outside air! The reason for this was that the supply air for the engineer's office in the basement was return air from the core of the building and, by overwhelming the reheat coils with 100% outside air, the building core temperature dropped a few degrees and, in turn, cooled the engineer's office a few degrees. Modeling the building with automatic control of outside air would not have produced an accurate simulation; in fact the building was modeled using an average outside air percentage of 70%. The very first output for the mainframe program simulation of this building showed a calculated energy use that was within 5% of the building's actual energy consumption.

In another downtown San Francisco high-rise, the chief engineer utilized a variety of electro-mechanical time clocks and "patch cords" to start and stop the building's various HVAC systems (he literally "plugged in" to whichever time clock he wanted a particular system to use). As he explained, he was then using the 7 a.m. to 6 p.m. time clock. Late night observation, however, backed up by review of building electrical demand recordings, revealed that he had inadvertently "patched" himself into the time clock set for 6 a.m. to 7 p.m., resulting in a 10% to 12% increase in the building's HVAC energy use. Modeling this building based on scheduling information obtained from "the horse's mouth" could never have provided an accurate simulation.

In larger buildings, not only are operational practices manyfold more dramatic in their effect than the effects of changes to the building envelope (which influence weather-related loads), but, we believe that the whole issue of weather data is greatly misunderstood in the industry. Some building simulation programs have been criticized in the past for not providing 8,760 hours of actual weather data for simulations. The well-known mainframe program developed by the Department of Energy (D.O.E.) and its various offspring provide 8,760 hours of simulation by means of (among others) a weather data source known as the TRY or "test reference year." Other programs, such as the TRACE program developed by the Trane Company, provide a weather data file consisting of an average 24-hour profile for each month of the year, for a total of 288 hours of simulation. In truth, there is little if any meaningful difference between these methods for two reasons. The first reason is that the "test reference year" is not an actual year's weather data. It is, in fact, an amalgam of 12 actual month-long "chunks' of data. These months of real data are selected for incorporation into the reference year by a process that effectively chooses the mean month out of the months of data available. Unless each weather data file is examined in detail, the user cannot be certain that "real' (whatever that means) weather extremes actually reside in the file or not. Furthermore, given the continuous nature of our solar system and the statistical difference between an "average" and the "mean," the true difference between 8,760 hours of simulation and 288 is difficult to discern, except in the run times of the various programs (which vary according to the number of hourly calculations that must be made). The second reason, which applies to new or existing buildings, relates to the purpose of performing a building simulation in the first place. The general thrust of any simulation is to project the future so as to make technical and economic decisions regarding building design or retrofit features. All of this presupposes that the weather that will actually occur in the future period under consideration (3 to 10 years generally) will be essentially equal to the weather data being used for the simulation. Since this cannot be known for a certainty, and the fact that weather-related factors are not dominant in determining energy use in the first place, any decision that would be influenced by the small effects in the calculations caused by the difference between 8,760 and 288 hours of simulation, would be a decision of doubtful wisdom at best.

Observational Surveys

As a result of experiences similar to the above, it has become our practice to perform two specific types of surveys in the buildings we study.

The first of these surveys is observational in nature and includes careful observation of the <u>functioning</u> of the building's temperature control systems—as opposed to simply reviewing the temperature control as-built drawings. We have found that frequently the controls were not installed as drawn, have been overridden (known as "auto-manual" control), or have simply failed in one fashion or another. This observational survey generally includes sample measurement of system operating parameters (supply air temperature, mixed air temperature, space discharge air temperature, etc.) as a means of observing the actual performance of the control system. The results of the inspection are frequently quite amazing!

The observational survey also regularly includes a "late night" tour of the facility and its HVAC systems to identify actual operating schedules (frequently at odds with what is reported by the operating engineers) and control system performance during this period. In one building surveyed, the control air compressor was off at night but the fans and pumps were still running—resulting in extreme overheating of the facility at night, which also made the chillers work hard in the morning to bring the building back down to temperature when the controls came back on! This late night survey is also invaluable in confirming the operating schedules for lighting systems, which are frequently under the control of the custodial crew.

Electrical Load Surveys
     The second type of survey we find essential is an electrical load survey. Where great accuracy is desired, such as in the modeling of large high-rise office buildings, every electrical panel and piece of equipment should have its instantaneous power draw measured. This can be done with a hand-held power factor meter, with the data recorded and entered into a spreadsheet developed just for this purpose (see Table 16-1, which is a sample output page from just such a spreadsheet). It should be noted that simply reading voltage and amperage is insufficiently accurate, as induction motors especially have very wide ranges of power factors (depending upon their loading) that can cause volt/amp readings to be in error by 50% or more when compared to true power draw. In addition to the instantaneous measurement of electrical loads, it is also important to look at specific large loads (chillers, elevators, computer rooms, etc.) over time, using a power-recording instrument. This instrument can also be used to observe the total power demand profile for the entire building if the building is small and time-of-day metering is not employed by the utility company. Frequently, particularly for large buildings, the utility company records the building's power demand over time (utilizing magnetic tape or bubble memory meters) and the information from these meters is almost always available from the utility company (see Figure 16-1, which shows a 24-hour plot of utility company demand interval records).

     As large buildings, even in cold climates, spend most of their time in a cooling mode of HVAC system operation, electrical energy use makes up the vast majority of the building's energy use. This being the case, it is important to compare the sum of the various instantaneous load measurements with the recorded peak demand for the building, as shown in Table 16-2. If the individual measurements

Table 16-1. Example of electrical load survey data tabulation.

POWER MEASUREMENT FORM

DATE:8/30/94   RECORDED BY: CDS &JPW

LOAD	TIME	AVE VOLTS	L-1	P.F.	L-2	P.F.	L-3	P.F.	KW	REMARKS
HV-1A	2:40P	278	3.7	0.23	3.8	0.33	3.4	0.29	0.9	East Campus Basement MCC
EF-1A	2:40P	278	1	0.43	1	0.54	0.9	0.48	0.4	East Campus Basement MCC
EF-3A	2:40P	278	1.1	0.49	1.2	0.59	1	0.58	0.5	East Campus Basement MCC
HV-2A	2:40P	278	2.1	0.33	2.1	0.44	1.9	0.41	0.7	East Campus Basement MCC
AC UNIT (OLD ICU)	2:40P	Not in Operation	0.00	0	0.00	0	0.00	0.0		Roof of Incinerator
P-1	2:40P	278 11.1	0.75	11.6	0.78	11.1	0.79	7.3		Heating Water Pump
P-4	2:40P	278 0.5	0.30	0.5	0.54	0.3	0.61	0.2		Heating Water Pump
P-5	2:40P	278 1.7	0.62	1.7	0.72	1.5	0.71	0.9		Heating Water Pump
CWB PUMP	2:40P	Not in Operation	0.00	0	0.00	0	0.00	0.0		Cold Water Booster Pump
CWB PUMP	2:40P	278	3.7	0.49	3.8	0.57	3.4	0.55	1.6	Cold Water Booster Pump
AC-1	2:12P	277	9.8	0.74	9.6	0.78	10.3	0.76	6.3	1st Floor MCC Panel
CHW PUMP	2:12P	277	1.5	0.46	1.6	0.37	1.4	0.44	0.5	1st Floor MCC Panel

Equipment	Time								Location
EXHAUST FAN	1:50P	276	0.7	0.50	0.7	0.50	0.8	0.41	0.3 3rd Floor Roof MCC
SUPPLY FAN	1:50P	276	16.5	0.83	15.4	0.87	15	0.81	10.8 On Roof
RETURN FAN	1:50P	276	5.8	0.83	5.5	0.83	5.6	0.80	3.8 On Roof
HEATING PUMP	1:50P	276	3.1	0.55	3.2	0.54	3.2	0.56	1.4 In Penthouse
EF-1 SURGERY	1:06P	277	2.7	0.54	2.7	0.63	2.4	0.58	1.3 Roof MCC
EF-2 SURGERY	1:06P	277	9.1	0.67	9	0.67	8.9	0.66	5.0 Roof MCC
EF-9	1:06P	277	11.4	0.37	1.5	0.49	1.2	0.50	1.5 Roof MCC
HV-2 STORAGE Area	1:06P	277	4.2	0.52	4.3	0.59	4	0.57	1.9 Roof MCC
HV-3 KITCHEN	1:06P	277	5.6	0.61	6	0.67	5.4	0.70	3.1 Roof MCC
HV-4	1:06P	277	8.4	0.82	9.1	0.85	8.4	0.87	6.1 Roof MCC
EF-4 RANGE HOOD	1:06P	Not in Operation	0.00	0	0.00	0	0.00	0.0	Roof MCC
EF-3 STORAGE	1:06P	277	3.3	0.49	3.4	0.56	3.1	0.53	1.4 Roof MCC
EF-7 ADMITTING	1:06P	277	3.1	0.68	3.2	0.73	2.9	0.74	1.8 Roof MCC
EF-6 TOILET EXT	1:06P	277	2.1	0.79	2.1	0.84	1.9	0.85	1.4 Roof MCC
EF-8 DINING RM	1:06P	277	1.6	0.60	1.7	0.67	1.4	0.73	0.9 Roof MCC
HV-1 ADMITTING	1:06P	277	3.1	0.70	3.3	0.72	3.1	0.75	1.9 Roof MCC

PAGE 1

Figure 16-1. Example of analysis of utility company demand interval records.

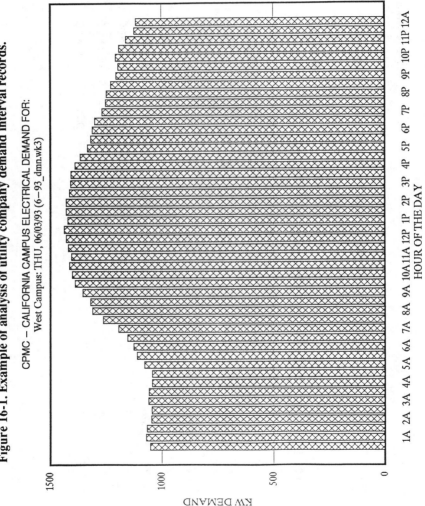

CPMC – CALIFORNIA CAMPUS ELECTRICAL DEMAND FOR:
West Campus: THU, 06/03/93 (6–93_dmn.wk3)

don't equal the total demand, then any attempt at modeling will fail. Furthermore, a building's energy use is determined by connected loads multiplied by hours of use. By utilizing the data from the operational survey and checking it against the record of electrical demand over time, a high level of confidence can be achieved as to the actual operating schedules of the various energy-using systems in the building.

Output Critique

One of the hardest things to do in performing a building simulation is to honestly critique the computer output. After spending hours or even days preparing the input data, it is easy to fall into the trap of believing that the output must be correct. However, as our mistakes prove to us, it is critically important to critique the computer output with a skeptical attitude. Three specific techniques are valuable in regard to critiquing simulation program output.

Annual Energy-Use Profile Comparison

The first technique is a gross, year-long evaluation of the modeled energy use in comparison to actual energy use. While the totals may agree, seasonal variations may not agree well with each other, indicating that weather influenced systems are not modeled well. Graphic comparison of modeled and actual energy use is most valuable in this evaluation, as can be seen in Figures 16-2 and 16-3. In addition, since computer simulations generally utilize weather data that are a composite of multiple years' data (including NOAA's "TRY" tapes as previously discussed), it is valuable to contrast the actual weather data for the year being modeled to the weather data employed in the simulation, as shown in Figure 16-4. When modeling a building using a year's worth of actual energy use for validation, it is more important that the modeled energy use vary according to the changes in the model weather for the same period rather than absolutely agree with the actual utility data being used for comparison. For example, if the model shows higher than actual electrical use for cooling in a given month and both the actual electrical use and actual temperatures are lower than the model, then this lends credence to the model and means the model is meaningful for evaluation of multiple future years' potential for energy savings.

Table 16-2.
Example of comparison of field-measured electrical loads to peak demand recorded by the utility company.

————CALCULATION SHEET————   DATE: 10/29/94
    (PRE-MODL)   INIT: JPW

PRELIMINARY MODEL OF CPMC, CA-WEST CAMPUS:

LOAD	PEAK KW	MIN KW	HR/YR	KWH
LIGHTING	492	246	5111	2514396
MISC PROCESS/OFFICE	212	85	4344	922661
AIR HANDLING UNITS				
SUPPLY FANS	243	243	8760	2128680
RETURN FANS	78	78	8760	683280
CHILLERS	340	170	2500	850000
CHW PUMP	30	30	3000	90000
CHW PUMP	20	20	429	8580
CW PUMP	25	25	3000	75000
CW PUMP	25	25	429	10725
CLG TOWER	15	15	3000	45000
CLG TOWER	15	15	429	6435
PROCESS USE				
DATA PROCESSING **	67	67	8760	586920
KITCHEN	107	107	7884	843588
CHERRY ST GARAGE	32	32	8760	280320
MISC FANS/PUMPS	32	32	8760	276904
MED AIR/VACUUM	30	30	8760	262800
TOTALS	1763	1220		9585288
ACTUAL	1662	1306		10140698
VARIATION	6.1%	-6.6%		-5.5%

————————ENERGY RESOURCE ASSOCIATES————————

**INCLUDES HVAC

Peak Load Comparison

The second of these techniques is to evaluate peak modeled loads against known values. From the utility company's data, the building's peak electrical demand is known for all seasons of the year. Generally, computer models will provide a monthly peak electrical demand for the various components of the model. By comparing the principal seasons (summer, fall/spring, and winter), it can be observed whether all of the loads measured during the survey found their way into the model and whether the seasonal modeling of cooling loads is correct. Furthermore, the building's peak cooling load is probably known from operating engineers' observations and/or operating logs, and this too, can be used as a scale of measure for evaluating the accuracy of the computer model. Again, the issue of the weather data employed for the simulation must be taken into consideration. Generally, all weather data used for simulations are

Figure 16-2.

Comparison of model and actual annual electrical use profile.

COURTHOUSE/ADMIN. BLDG. ELECTRICITY

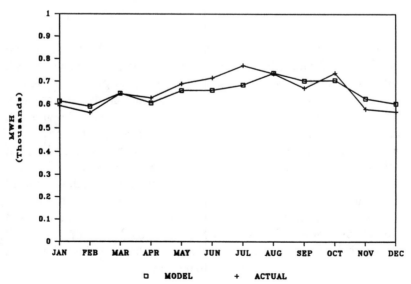

Figure 16-3.
Comparison of model and actual annual gas use profile.

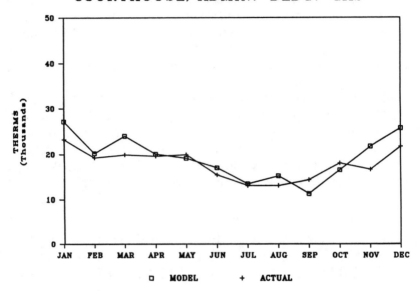

missing the hottest and coldest days of the year. Accordingly, the actual demand data used for comparison would best be selected as a day experiencing the same, or nearly the same, temperature extremes as present in the weather data used for simulation. Interestingly enough, one building we modeled had one of its chillers fail and was short of capacity to support anything close to a "design" day. As a part of assessing the comfort "risk" caused by the failed chiller, the mainframe simulation output was reviewed and we identified the ambient temperature at which the simulation would predict "losing" the building on a hot day. In fact, within a few weeks of completing the modeling process, an unseasonably hot day was encountered with a peak temperature exceeding our predicted "lose the building" temperature by a few degrees. Indeed, the chief engineer reported that he had "lost" the building on that one day.

Figure 16-4.
Comparison of model and actual weather data.

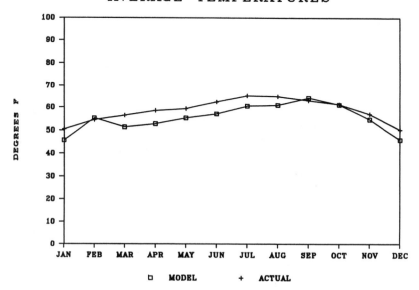

AVERAGE TEMPERATURES

□   MODEL   +   ACTUAL

Detailed Output Analysis

The third technique is primarily oriented towards evaluation of energy retrofit models. In order to develop savings estimates for energy retrofit measures under consideration, the retrofit is modeled and then contrasted with the original model, thus showing the savings that might be achieved. Since it is very easy to make small errors in editing the input data for a computer model and cause an unintended result, a useful quality control technique has been to analyze the computer model in detail (by functional use, i.e., lighting, cooling, fans, pumps, etc.) and develop a specific figure for the savings estimated for each retrofit in each functional use area. As can be seen in Table 16-3, a very detailed analysis of the output from a mainframe proprietary computer model is possible. The analysis allows a "plausibility" check of the savings from a particular retrofit.

Table 16-3. Example of detailed analysis of output from "mainframe" simulation.

CALCULATION SHEET
(TRC_ANL1)

DATE: 3/24/89
INIT: JPW

ANALYSIS OF TRACE RUN/ALTERNATES TO DETERMINE ENERGY SAVINGS:
ECM #   1 & 2
TITLE   ADMN BLDG PENTHOUSE AND BASEMENT DOUBLE DUCT TO VAV

EQUIPMENT	ENERGY	BASE	COMP TO	ECM USE	DELTA	% REDUCTION
CHILLER 1	KWH	442052	442052	430574	11478	2.6
CHLR 1 AUX	KWH	157787	157787	155471	2316	1.5
CHILLER 2	KWH	390069	390069	337348	52721	13.5
CHLR 2 AUX	KWH	70208	70208	64355	5853	8.3
CHILLER 3	KWH	44762	44762	25963	18799	42.0
CHLR 3 AUX	KWH	38332	38332	24441	13891	36.2
BOILER	THERMS	191586	191586	137840	53746	28.1
BOILER AUX	KWH	65831	65831	60051	5780	8.8
SYS 1 SF	KWH	397881	397881	129122	268759	67.5
SYS 1 RF	KWH	183582	183582	59574	154008	67.5
SYS 1 EF	KWH	13690	13690	5397	8293	60.6

SYS 2 SF	KWH	46228	46228	46228	0	0.0
SYS 3 SF	KWH	38082	38082	38082	0	0.0
SYS 4 SF	KWH	147921	147921	147921	0	0.0
SYS 5 SF	KWH	48861	48861	48861	0	0.0
SYS 5 RF	KWH	1972	1972	1972	0	0.0
SYS 6 SF	KWH	16701	16701	16701	0	0.0
SYS 6 RF	KWH	3336	3336	3336	0	0.0
SYS 7 SF	KWH	83202	83202	83202	0	0.0
SYS 8 SF	KWH	67196	67196	67196	0	0.0
LIGHTS	KWH	1827783	1827783	1827783	0	0.0
BASE ELEC	KWH	3652192	3652192	'652192	0	0.0
BASE GAS	THERMS	39909	39909	39909	0	0.0

TOTAL ELECTRIC SAVINGS: 511898 KWH
TOTAL GAS SAVINGS: 53746 THERMS

ENERGY RESOURCE ASSOCIATES

For example, if a variable-air-volume retrofit is under consideration, it is possible to develop a specific estimate for the savings to be achieved by the fan alone. This savings can then be compared to the original energy used by the fan and the plausibility thereof evaluated. If a simple inlet vane conversion is anticipated and the system operates a single shift per day during weekdays, a savings figure in the neighbor hood of 30% to 40% might be anticipated on a "rule of thumb" basis. If the detailed analysis indicates a savings of 70% or 80%, then review of the model input is warranted to determine the error in the input or determine the reason that a savings figure much higher than the engineer's "rule of thumb" is reasonable. For example, perhaps the system does, after all, operate on a 24-hour per day basis or was grossly oversized and will experience very low loads compared to its installed capacity for most of its operating hours. In any event, when the savings vary greatly from that which is "plausible," it indicates either an error in the modeling or an error in the plausibility logic—either of which should be determined before using the savings numbers generated by the model.

It is theoretically possible to create a "perfect" model in which every small unique thermal zone in a building responds to weather inputs virtually the same as the actual building. However, the practicality of such modeling is doubtful, as the engineering costs to prepare such a model may actually exceed the value to be created by the modeling process, particularly in smaller buildings. As a result, even the best modeling tools and reasonably constructed models will be limited in their ability to predict the effect of retrofit measures. Therefore, in some cases, it is an appropriate engineering step to de-rate or discount the savings figures for engineering conservatism (see Table 16-4 for example). A good example of this is the fact that many computer models that utilize hourly heating and cooling load calculations as part of their modeling (not all do, as we shall see below) are unable, without laborious and extensive micro-zoning of the model, to avoid the sharing of internal heat gain with external zones needing heating and thus underestimate the actual heating requirements of the building. Similarly, tall buildings in central city locations often have large vertical exterior zones, part of which need cooling and part of which need heating at any given time, primarily due to solar exposure and shading from adjacent buildings. These perimeter systems can be difficult to model and sometimes will show optimistic results from even the most conservative attempts at modeling—thus necessitating an engineering discounting of savings. The bottom line here is that even the best models still have limits to their capabilities—even when using the most complex simulation programs available!

Plausibility Check

Finally, by summing all the savings for all retrofits, a gross plausibility check can be performed, based on engineering judgement regarding whole building energy-use levels that are reasonable for the type of building being evaluated. This is a gross measure, but it is an excellent <u>final</u> check on the entire process, as shown in Table 16-5. Even such a simple check can be effective in catching unreasonable optimism in energy savings estimates that may have slipped through all the other quality control measures in this very complex process of building simulation. Had such a macro check been part of the project documentation associated with the project mentioned in the introduction, that energy services company would not have the problem they currently face.

SIMULATION TOOLS

It is likely that a wide range of opinion exists in the energy engineering field as to what constitutes "building energy simulation." Our view is a rather broad one and encompasses a wide range of calculational strategies as being appropriate to specific project goals and project environments.

"Mainframe" Programs

The high end of the practice are programs that have traditionally run on mainframe (or mini) computers. Both proprietary and public-domain programs are in common use, and include such programs as DOE-2, TRACE, ESP-II and others. The availability of such programs to run on high-end personal computers has become fairly

Table 16-4. Example of conservative derating of energy savings from a simulation model.

————CALCULATION SHEET————
(WCT-ECM2)

DATE: 10/29/94
INIT: JPW

ANALYSIS OF TRACE RUN/ALTERNATES TO DETERMINE ENERGY SAVINGS:
ECM #   2
TITLE   WEST CAMPUS, BUILDING AUTOMATION SYSTEM

EQUIPMENT	ENERGY UNITS	BASE MODEL USE	COMPARE ECM TO: - BASE	ECM MODEL = USE	CHANGE IN USE	% REDUCTION FROM BASE USE
PRIMARY HTG	KWH	728982	728982 -	706808 =	22174	3.0
	THERMS	458035	458035 -	409214 =	48821	10.7
PRIMARY CLG						
COMPRESSOR	KWH	673322	673322 -	382807 =	290514	43.1
TOWER FANS	KWH	47730	47730 -	28865 =	18864	39.5
COND PUMP	KWH	137788	137788 -	81035 =	56753	41.2
OTHER ACCES	KWH	8122	8122 -	4829 =	3292	40.5

AUXILIARY						
SUPPLY FANS	KWH	3042355	−	2971744 =	70611	2.3
CIRC PUMPS	KWH	310135	−	218513 =	91622	29.5
BASE UTIL.	KWH	1850988	−	1850988 =	0	0.0
LIGHTING	KWH	2883474	−	2883474 =	0	0,0
RECEPTACLE	KWH	986380	−	986380 =	0	0.0
DHW & PROCESS	THERMS	347034	−	347034 =	0	0,0
COGENERATION	KWH	0	−	0 =	0	100.0
	THERMS	0	−	0 =	0	100.0

TOTAL SAVINGS: 553831 KWH   48821 THERMS
TECHNICAL PLAUSIBILITY FACTOR: 0.95   0.85

NET SAVINGS: 526139 KWH   41498 THERMS

———— ENERGY RESOURCE ASSOCIATES ————

Table 16-5. Example of "gross" or "overview" check of savings calculations.

———— CALCULATION SHEET ————   DATE: 4/5/89
(CABC_SUM)   INIT : JPW

ENERGY SAVINGS SUMMARY

ENERGY CONSERVATION MEASURE;	KWH	THERMS	$$$
1. ADMIN PENTHOUSE DOUBLE DUCT TO VAV &			
2. ADMIN BASEMENT DOUBLE DUCT TO VAV	511898	53746	$62,246
3. ADMIN COURTROOM MULTIZONES TO VAV	37124	3589	$4,415
4. CONVERT JAIL MULTIZONES TO VAV &	53579	0	
7. LARGE COURTHOUSE MULTIZONES TO VAV	127761	36682	
SUB TOTAL	181340	36682	$27,696
5. SUPERVISOR'S AHU CONTROL MOD	12372	332	$1,195
6. LIGHTING RETROFIT	448987	–1946	$38,888
8. COURTHOUSE SMALL MZ'S TO VAV &			
9. COURTHOUSE SMALL MZ TO VAV	148703	43771	$27,093

		BTU/SF/YR			
10.	SUMMER STEAM SHUT-DOWN		0	10287	$3,292
11.	ENERGY MANAGEMENT COMPUTER		74974	6190	
	63146		0		
	SUB TOTAL		138120	6190	$14,135
12.	VARIABLE FLOW CHILLED WATER		15647	0	$1,377
	TOTAL (EXCL ECM#12)	43680	1478544	152561	$178,960
	PLAUSIBILITY FACTOR		0.95	0.95	
	NET SAVINGS	41496	1404617	145018	$170,012
	EXISTING CONSUMPTION	104035	7896022	214272	$763,417
	PERCENT REDUCTIONS	39.9	17.8	67.7	22.3
	RETROFIT BTU/SF/YR	60355			

1. ELECTRICITY AVERAGE UNIT COST FOR 12 MO. ENDING OCT '88 WAS $0.0796/KWH, PLUS APPROX 10% PG&E RATE INCREASE IN JAN '89 EQUALS $0.088/KWH USED ABOVE.
2. NATURAL GAS UNIT COST USED IS $0.32/THERM.

———— ENERGY RESOURCE ASSOCIATES ————

commonplace. In general these programs have similar, if not common, ancestry and are founded in hourly heating and cooling load calculations that are then applied to the HVAC systems and equipment described to the program. These types of programs are powerful simulation tools, allowing for detailed input of both the envelope and the lighting and HVAC systems in the building, and produce excellent results (see Figures 16-2 and 16-3). Also, these programs provide extensive output data for use in output critique. While very powerful, these programs require significant engineering labor to prepare the data necessary for input (often 40 to 80 engineering labor hours, even for fairly straightforward models) and are sometimes too costly for use on smaller buildings or for use in the qualification of sales prospects in the energy retrofit business. To meet the need for less costly simulation methods, we developed some spreadsheet-based simulation tools that have proven to be very effective.

Complex Spreadsheet Simulation Tool

Another possible simulation tool is a complex, automated spreadsheet that allows time-related loads to be scheduled by hour, by three day types (Weekday, Saturday and Sunday/Holiday), by type of energy used, or by type of functional energy use (cooling, fans, lighting, etc.). Too, the calendar of day types for the model year can be customized to cover most any situation. With respect to weather-related loads, this model takes a totally different approach than mainframe programs. In this case, the program accepts peak loads as inputs and distributes the loading over the period of a year according to the differential between the modeled ambient temperature and user-input "no-load" temperatures for heating and cooling. Other variables include heating and cooling lockout temperatures, minimum loads, and daily and seasonal operating schedules. The model calculates hourly ambient temperatures for application of the loads by using a near-sinusoidal model and varying the temperature up or down from the average temperature by half the average daily range. The model utilizes as input degree-days and average daily range by month, or average maximum and minimum temperatures by month. The model provides hourly heating and cooling loads, and hourly time-related loads, for typical day types each month.

As can be concluded from observation of Figures 16-5 and 16-6, this modeling tool can produce simulations of high accuracy and requires only a few hours for input generation and model runs. In addition, because there is great control over the model, many different retrofit measures can be modeled and custom simulations can be produced by modifying the code or extracting output from the base building model and performing subsequent calculations thereon. This tool is most effective on smaller or simpler buildings, where a high level of confidence in energy-savings figures is desired but engineering costs must be kept to a minimum, and is finding favor among contractors, energy service companies and utility companies.

Simple Spreadsheet Simulation Tool
Another possible simulation tool is a one-page simulation spreadsheet. Its purpose was to provide an extremely quick and inexpensive simulation tool for use where limited accuracy is accept-

Figure 16-5. Example of complex spreadsheet simulation results.

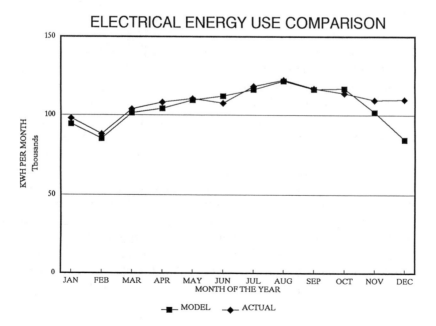

Figure 16-6. Example of complex spreadsheet simulation results.

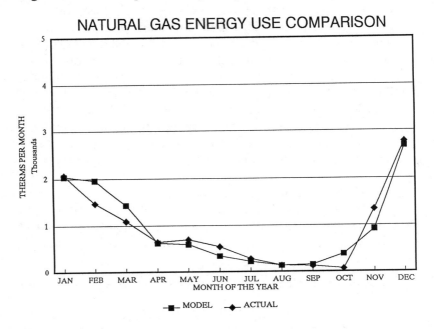

able and simulation costs are of greater importance than accuracy. Two versions of this model exist, one for HVAC systems that mix heating and cooling (e.g., terminal reheat) and one for non-mixing systems. As shown in Table 16-6, this simulation tool has very simplistic input and basically views a building as having lighting, heating, cooling, HVAC accessories, domestic hot water, and two types of miscellaneous energy use (electrical and heating fuel). Inputs are generally in units per square feet (e.g., lighting input is in watts per square foot) and percentage of operating hours. In addition, provision is made for reduced summer operation (primarily for schools) and "off hours" loads in all functional areas. Time-related loads are calculated based on "hours on" times input loads, similar to the spreadsheet described above, without the ability to customize day types or the annual calendar. Weather-related loads assume a linear, directly proportional relationship with degree-days, which are input to the spreadsheet.

Table 16-6. Example of input and output from simple spreadsheet simulation.

SWEENY COMMUNITY HOSPITAL ENERGY USE TEMPLATE
FOR NON-MIXING SYSTEMS

ENERGY UNITS>>>	UNIT #	TYPE
	1	ELEC, KWH
	2	GAS, MBTU
	3	STEAM,MBTU
	4	CHW, MBTU

INPUT BY [M. MUNIZ]   DATE [9/7/89]

AREA NAME [EMERGENCY]   SYS NO. [8-2]   OCCUPIED SQ. FT. [3200]   YEAR BUILT [1974]

Input parameters

	LIGHTING	COOLING	HEATING	CLG/HTG ACCESSORIES	DOM. HW OFFICE	DOM. HW RESID.	MISCELLANEOUS A	MISCELLANEOUS B
:NRG UNIT >	1	1	2	1	3	3	1	2
	2.1 :WATTS/SF	350 :SF/TON	20 :KW/TON	0.2 :SF/PERSON	150 :SF/PERSON	200 :WATT/SF	0.5 :MBH CAPAC	
		1 :EQU. FCTR	0.3 :EQU. FCTR	0 :GAL/PER/DA	0 :GAL/PER/DA	0 :% OCCUP.	0 :% OCCUP.	0 :% OCCUP.
		0.1 :EQU. FCTR	1.2 :EQU. FCTR	1 :TEMP FCTR	1 :TEMP FCTR	1 :UNOCC LD %	60 :UNOCC LD %	60 :UNOCC LD %
% HRS OCC.>	80	60	60					
:UNOCC LD%>	80	60	100					
:MIN LOAD %>		100	100 :HTG USE %	100 :% SUMMER LD%	100 :% SUMMER LD%	20 :% SUMMER	20 :% SUMMER	100 :% SUMMER
% SUMMER >								
:assumed:				140 :HWS TEMP	140 :HWS TEMP			
				60 :CWS TEMP	60 :CWS TEMP			

:KW TOTAL >	7 :TONS	9 :MBTU'S	64 :KW	1.8 :OCCUPANCY	21 :OCCUPANCY	16 :KW	2	

USAGE

USAGE	UNITS NRG	SEPTEMBER	OCTOBER	NOVEMBER	DECEMBER	JANUARY	FEBRUARY	MARCH	APRIL	MAY	JUNE	JULY	AUGUST	TOTALS
LIGHTING	ELEC, KWH:	4,645	4,800	4,645	4,800	4,800	4,335	4,800	4,645	4,800	4,645	4,800	4,800	56513
COOLING	ELEC, KWH:	4,627	2,694	1,451	74	42	672	545	2,232	3,575	5,256	6,857	5,798	33843
HEATING	GAS, MBTU:	0	0	2,905	17,142	12,876	7,456	3,910	0	0	0	0	0	44289
CLG/HTG ACCESSORIES	ELEC, KWH:	1,317	1,360	1,317	1,360	1,360	1,229	1,360	1,317	1,360	1,317	1,360	1,360	16018
DOM. HW OFFICE	STEAM,MBTU:	0	0	0	0	0	0	0	0	0	0	0	0	
DOM. HW RESID.	STEAM,MBTU:	0	0	0	0	0	0	0	0	0	0	0	0	
MISCELLANEOUS A	ELEC, KWH:	783	809	783	809	809	731	809	783	809	783	809	809	9531
MISCELLANEOUS B	GAS, MBTU:	0	0	0	0	0	0	0	0	0	0	0	0	
EMERGENCY														
SYS. NO. 8-2	ELEC, KWH:	11,371	9,664	8,196	7,064	7,012	6,767	7,515	8,977	10,545	12,000	13,826	12,767	115905
	GAS, MBTU:	0	0	2,905	17,142	12,876	7,456	3,910	0	0	0	0	0	44289
OCC. SQ FT 3200	STM, MBTU:	0	0	0	0	0	0	0	0	0	0	0	0	
	CHW, MBTU:	0	0	0	0	0	0	0	0	0	0	0	0	

TOTAL COMBINED BTU/SF   137,624

This model was developed to simulate a college campus of more than 100 buildings (all of which had fairly simple HVAC systems)) using one model per building. This tool was also used to model a small community hospital that had a very large number of very different HVAC systems. This model was used to simulate each of the hospital's HVAC systems individually with the modeling accuracy results as shown in Figure 16-7. Considering the relatively small amount of engineering effort required for modeling, the results were excellent. Another appropriate and attractive use of this spreadsheet simulation tool would be as a first-order conservation assessment tool in the energy conservation sales process.

BUILDING SIMULATION AND ENERGY SERVICES

In the last ten years or so, a mini-industry has formed that has traditionally been referred to as the "energy services" industry. Recently, the term "demand-side management" has also been applied to

Figure 16-7. Example of simple spreadsheet simulation results.

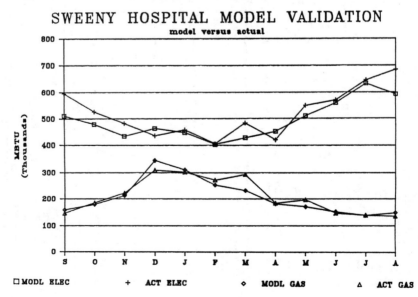

SWEENY HOSPITAL MODEL VALIDATION
model versus actual

☐ MODL ELEC   + ACT ELEC   ◇ MODL GAS   △ ACT GAS

this business. What is essential to this industry is the business proposition of retrofitting an owner's building at essentially no initial cost to the owner (financing is provided by the energy services company or a third party) and guaranteeing in some fashion that the utility cost avoided by the project will equal or exceed the cost of the project (debt service plus any other ongoing costs such as project management or maintenance). Unfortunately for some projects done in this industry, the sales people involved viewed the business proposition as simply a way to make their job easier and they exhorted their technical staffs to generate savings calculations that would support a high dollar value for their projects. This is unfortunate, and even frightening, because savings figures so generated are difficult if not impossible to achieve in reality and, if the guarantee offered is reputable, it must then come into play to cover the savings shortfall that must necessarily occur. In a most dramatic example, one energy services company with which the author has worked had the unpleasant experience of having a sales engineer substitute his own savings calculations for those generated by the computer model. The result of this was a guarantee of natural gas savings on one project that actually exceeded the natural gas consumption of the building. Needless to say, management failed to properly consider the plausibility of such a proposition and approved the project for funding—and wound up funding the annual savings "shortfall" to the tune of more than $100,000 per year (not to mention destroying their relationship with the building owner). The use of cost-effective and accurate tools and methods of building simulation is an essential part of identifying and implementing successful energy services or demand-side management projects.

While it is a fascinating and complex engineering tool, the fundamental value of computer simulation of buildings is that it forces a quality-enhancing step in the analytical process. This step is essentially a systematic confirmation of the engineer's knowledge of where and how energy is being used in a building. If the modeling step is done and done well, it is difficult to make "off target" recommendations for specific types of retrofits or "off target" estimates of savings. With such a high level of confidence established on the technical side of a project, the assessment and mitigation of project performance risk can rightly be performed on the financial side of the project evaluation, resulting in a very high probability of success for energy retrofit and demand-side management projects.

CONCLUSION

As has been discussed herein, tremendous opportunities exist to refine, improve and automate calculational and other procedures which are part of the energy auditing process. Caution is advised, however, regarding the too-rapid introduction of new computer-based methodologies into the practice of energy auditing, as it is all-too-easy to get lost in the morass of technology and lose direction and momentum in the process of an audit. We caution adding new procedures and tools slowly, in an evolutionary fashion, so that valid use thereof can be established and confirmed one at a time.

REFERENCES:

"Energy Conservation With Comfort," 2nd Edition, Honeywell, 1979

"Energy Conservation Control" (EFACT Manual), Johnson Controls, 1982

"Practical Experience in Achieving High Levels of Accuracy in Energy Simulations of Existing Buildings," ASHRAE Transactions: Symposia, AN-92-1-2

"Computerized Building Simulation... A DSM Strategy?" Globalcon '94 Proceedings, 1994

"TRACE 600 User's Manual," The Trane Company, 1992

17

Preparing for New Options: Innovative Tactics and Technologies In a Competitive Utility Environment

ABSTRACT

A new age is dawning for lower-cost energy use and supply. The deregulation of the electric industry is creating new pricing structures that will change how we calculate the payback of alternatives for cutting the cost of energy in our facilities.

Energy users can help themselves navigate these choices. By understanding the concepts inherent in deregulation, learning to use new energy tools, influencing the deregulation process, applying new technologies, and (most important of all) by being as creative as possible, smart users will grasp new options for energy cost reductions.

AT&T VS. MCI: A PARADIGM

To better understand how electric utilities will be transformed, think about how long-distance phone service has changed. When AT&T was forced by a federal judge to divest itself of many of its divisions, long-distance was separated from local service, and new providers such as MCI and Sprint became household names.

Presented at 20th World Energy Engineering Congress by Lindsay Audin, CEM, CLEP, IES, president, Energywiz, Inc., One Everett Avenue, Ossining, N.Y. 10562; energywiz@aol.com

While, after 12 years, deregulation of that industry is not yet complete (local service is still generally a monopoly), during that time long-distance use has nearly quadrupled, while the average price of a long-distance call fell by more than 50%. To satisfy consumer demand for communication services, a vast new array of technologies was also born. How many of us anticipated the proliferation of fax machines, cellular phones, pagers, and on-line services that would result from a single court order? While we can also expect the cost of power to fall, the future of electricity similarly holds much more than price reductions.

FACTORS IMPACTING POWER PRICES

Electricity is generated by utilities and independent power producers (IPPs), both regulated to some degree by state public utility commissions (PUCs), and then transmitted through high-tension lines criss-crossing North America in a giant network. These lines are owned by utilities and regulated by the Federal Energy Regulatory Commission (FERC). Once voltage is stepped down at substations, power is distributed through local utility-owned lines and meters regulated by PUCs.

With the exception of municipal utilities (which are controlled by local governments), PUCs determine how to distribute these costs to end user classes. Most of our bills break out only charges for electric consumption and demand (and perhaps a fuel adjustment), but the true cost of power includes many other components, including transmission, distribution, and a variety of ancillary services (such as voltage support, spinning reserve, and load following). Bills may also include taxes, social programs, and other charges that are not apparent to end users.

To develop the prices we pay, the PUCs apply a standard based on the utility's costs for providing a service, plus a guaranteed rate-of-return to ensure a ready supply of investment capital. All of these costs and profits are "bundled" together to create tariff pricing. While theory dictates that charges should based on the true cost-of-service, politics and other pressures often result in cross subsidies in which one rate class (e.g., industrial) is charged more to contain prices charged to another (e.g., residential).

While the electric rates we pay are controlled by PUC tariffs, utilities and IPPs buy and sell electricity among themselves, and such wholesale prices vary with time, climate, power plant outages, fuel prices, and other factors. The base cost of power seen by a utility is therefore a mix of its own generating costs and the price it pays for electricity delivered to it through the transmission system from other power providers. This base cost is subject to commodity market conditions usually not visible to end users. It is increasingly being influenced by factors such as commodity trading techniques (e.g., futures and financing plans), ways to adjust user load profiles (such as real-time pricing), and transmission system constraints (that can drive prices up to the highest local generating cost).

THREE GENERAL RELATIONSHIPS

There are three general relationships that clarify how new techniques both interact and can be applied to control energy pricing. They are:

- time - load - price
- generation - transmission - natural gas options
- load shaping - financing methods - user technologies.

Time, Load, and Price

As deregulated retail power prices begin to vary like those at the wholesale level, they will become more time-sensitive. Since one's demand for power generally changes during the day and the week (and by season) we can expect the average price to also change with time and use (unless controlled by other factors, discussed below). As a result, load profile shapes (i.e., a graph of power versus time) will influence pricing, with flat profiles generally having the lowest average cost. Utilities generally typify such patterns via load factor, defined as average demand divided by peak demand. A high load factor would indicate a flattened profile while low load factors would occur where a peak demand is relatively brief, and is surrounded by much lower demand during the rest of the day. While most non-industrial building power demand varies with time, it usually does so in predictable patterns. Knowing the shape of your typical load profile can often reveal ways to cut the present and future cost—and price—of power, while also helping your power supplier offer the best and most secure pricing.

Like Sprint's "dime-a-minute" long distance rate, marketers will likely offer highly simplified rates that smooth out such time-based price variations, but subscribing to such options will not yield the lowest average power costs. Rates that vary widely over time may provide the lowest average price, and techniques that cut, level, or shift peak demand will help reduce those prices.

Transmission, Generation, and Natural Gas Options

In some cities and states, peak loads exceed transmission capacity many hours each year. When low-cost power can't be brought in, prices could be bid up to the highest local generating costs. A good example of such constraints appeared during the early hot spell of June 1997 in the PJM (Pennsylvania-New Jersey-Maryland) power pool. While daily bulk wholesale generation prices (which make up 40% to 65% of most bills) generally don't vary from one end of the pool to another by more than $.01/kWh, June saw variations exceeding $.13/kWh when transmission constraints blocked cheap power from reaching high-cost areas[1].

In some urban areas with older power systems (such as New York and San Diego), transmission constraints could yield similar results. Such areas with constraints are sometimes called "load pockets" during the period of constraint (which may exceed 1000 hours a year). In the United Kingdom, which uses a national power pool supplied by deregulated generators, power suppliers have also found ways to "game" the system to purposely congest transmission, thereby driving up the price of their product[2].

To address such possibilities, energy marketers have begun examining and/or promoting new local generation (or cogeneration) facilities, either at customer-owned sites or through re-powering of obsolete utility plants inside the load pockets. Natural gas generators with very low emission levels have become quite cost-competitive for both peak shaving and as base load power, opening the door to competition during transmission constraints. Similarly, a variety of technologies (discussed later in this chapter) exist to reinforce existing transmission systems. A recent study[3] found that a small investment toward improving transmission capacity in California could have a major impact on limiting power prices.

One way to shift peak demand is to substitute natural gas for electricity during peak pricing periods. As will be discussed below, a variety of technologies exist for using gas to directly provide horsepower, cooling, air compression and other power-intensive needs. Such convertibility will create truly interruptible energy rates, allowing clever end users to contract for both interruptible gas and power, attaining the lowest possible energy prices. Under these circumstances, transmission, generation, and natural gas options will compete with each other, driving all prices down over time.

Load shaping, Financing, and User Technologies

A variety of choices are emerging to configure loads in advantageous ways. While most have existed in one form or another, deregulation will allow marketers to help end users gather—and segregate—their loads more readily through metering and contractual means.

Coincident metering has been used to cut the average cost for power at facilities with many meters on different accounts held by one customer. At Columbia University in New York, for example, several dozen accounts existed on one property, the result of gradual expansion without attention to energy costs. Each account peaked at a different time, but (due to tariff construction) the sum of the bills was the same as though all buildings had peaked at the same time. By combining the accounts under one master demand meter, the average cost of power was cut by over 10%. Such combination will become easier under deregulation as usage (for the power commodity, as versus transmission and distribution) can be contracted under one account.

Load isolation, while not favored by utilities, can allow an end user to segregate loads that would be cheaper under utility tariff-based power than under market-based power, thereby reducing the average cost for all loads.

Demand cooperatives (already promoted by Planergy, Inc.[4]) are a way for end users to work together (typically through an organizing vendor) to obtain lower utility rates by pooling their interruptible loads and agreeing to curtail them when requested by the utility. Creating such a cooperative eases the difficulty of ensuring that any one load must be interrupted by allowing the demand to apply to a group of loads, only a few of which would be interrupted at any one time.

District-wide systems serve multiple customers with (for example) chilled water from a central facility that, in effect, transfers many individual electric chiller loads (which lower load factor) to a high load factor central facility that uses both electric and gas-driven units to minimize total cooling costs. Existing chillers may remain, but are bypassed by a connection to a common chilled water loop serving numerous facilities.

Bill Consolidation allows many accounts held by one customer to be gathered for both coincident metering and attainment of cheaper energy block load rates previously beyond any one account.

Aggregation involves the gathering of different customer accounts through a third party for purposes of bulk power purchasing, coincident metering, bill consolidation, transmission capacity reservation, and expert load analysis. Aggregators work along the same lines as MCI, buying co-ops, credit card handlers, and other organizations that compete for the privilege of bringing many end users together. All try to provide lower prices through bulk purchasing and handling. Present-day utility customers are, in effect, already "aggregated" into rate classes, but only to the point of developing a rate based on an assumed typical load profile. Each of these options allows for re-shaping of loads (as portrayed to utilities and regulators) without major alterations to end user facilities.

In similar ways, financial tools exist that can cut or levelize costs more easily than with engineering. Both marketers and financial firms will (or already do) provide access to futures, options, and payment plans that ensure prices do not exceed a predetermined level. While a full discussion of financial instruments is beyond the scope of this paper, end users need to understand the following basic concepts.

- Electric futures (which are available on the West coast, and soon in the East) are tradeable contracts for monthly blocks of firm power, purchased in advance of need, that will later be provided during normal business hours at a predetermined price. They are, in effect, negotiable promises for supply of power, though there is not necessarily a guarantee of delivery unless accompanied by a secure transmission arrangement.

- Options (typically in the form of "calls" and "puts") are contracts between suppliers and marketers (and thus end users) that allow a user of power to know that he can "call" for power and be ensured supply, or that a seller of power can be ensured a buyer when he "puts" out an offer to sell, both at pre-determined prices.

- Payment plans have been offered to low-income customers by utilities for years. Now, however, marketers are ensuring levelized (or pre-determined) monthly bills (not just pricing) through risk management techniques that involve financial and load analysis tools. While some monthly bills may be higher than in prior years, other months will be lower, and the annual total will be confined to a narrow range. Such plans are often complicated, requiring careful analysis.

- Tolling allows an end user (or his designee) to provide a generator with boiler fuel (typically natural gas) in trade for electricity at negotiated, non-tariff, rates. This process is common among power marketers also trading natural gas, and is used at times that utilities have excess generating capacity that can provide power into another utility's territory.

While all of these options provide end users with choices for controlling energy pricing, the impact of many of them can be maximized when used in cooperation with new technologies. Just as we have seen an explosion of choice in communications, it is likely that the future will see a variety of ways to create, store, and manage power. Many are already being offered, or are in the prototype stages of development. Following are several being supplied or considered by both energy marketers and equipment vendors.

Metering. Marketers are using more sophisticated power metering (e.g., ENet, MC³) as a sales tool to ensure that their clients' power costs are minimized through a variety of methods. The new standard in metering involves wireless communications, non-invasive wiring, and hourly, quarter-hourly, or real-time monitoring. Human meter readers, obsolete for many years, will likely disappear as more cost-effective systems are installed. Such systems will allow better load control, more accurate power nominations, better pricing, and a reduction in theft and tampering (a major problem at the residential and small commercial level).

Software. Computerized building simulations and load analyses have taken on new prominence as tools for predicting and flattening load profiles. Analyzing short intervals (1/4 hour) has become essential to maximize savings through tighter load control. Names and acronyms such as PEDA, RBOSS, and PowerManager will be heard quite often as marketers offer new services to cut electric bills and power pricing.

Energy Management Systems (EMS). As responding to real-time or market-based pricing signals becomes a common way to attain savings, an EMS takes on new importance for controlling variable loads, such as fans, pumps, DHW heaters and chillers. When tied into the metering and software tools mentioned above, the load managing power of an EMS can be greatly enhanced.

Power Storage Devices. While many power practitioners and regulators continue to assume that power cannot be cost-effectively stored, flywheel power storage units are now in use as uninterruptible power systems (UPS) to supply "clean" power, and small (2 kWh) units act as backup power for cable TV systems. Larger units (over 12 kWh) are being prototyped as peak shifters/ shavers for buildings. Chemical battery technology has advanced considerably, and may play a part, as well. Depending on how stranded costs are collected, a storage system that gets "filled" with cheap off-peak power at night (when there are no transmission constraints) and "empties" to flatten peak loads during the day, could be an instant moneymaker for a smart marketer and/or end user.

Distributed Generation. The ability to generate power in a pinch has always been useful. New small (<100 kilowatt) modular gas-fired turbine generators[5] can also cut billing for peak demand by operating in parallel with utility power, or (to avoid backup charges) by feeding dedicated loads, while acting as backup power when needed by the utility's transmission and distribution system, thereby increasing overall system reliability. Using ceramics and limiting movement to a single part, the Capstone Turbine is an intriguing option, while the ONSI 200 kW fuel cell (and its soon-to-be seen competitors) has already racked up an impressive operating record. Once prices on these devices come down, watch marketers and end users grab them up as a way to minimize stranded cost payments.

Transmission and Distribution (T&D) Networks. Since power is cheaper only when transmission is available to move it, pressure for new or more robust transmission will increase once large price differentials occur between adjacent areas. Thyristor-based switching of high-voltage loads on T&D systems can raise the effective capacity of existing transmission lines. Such options are among a family of Flexible AC Transmission System (FACTS) improvements under development or deployment. Even new types of underground high-voltage cables[6], designed for use in transmission-constrained urban areas, are being rolled out to meet the expected demand for beefed up transmission.

Gas-Powered Motor Drives. Natural gas-driven devices, such as gas engine-powered air compressors and chillers, have replaced electric motor-driven units in industrial and commercial facilities, cutting their peak demand.

Advanced HVAC Systems. Chemical desiccants dehumidify outside air using natural gas, thereby cutting peak electric chiller loads. This process is already common in new buildings with large outside air loads (e.g., hospitals) and industrial processes (such as air compression). For smaller facilities configured around rooftop units, Entergy has been offering a super-high-efficiency replacement unit which takes advantage of several refrigeration engineering innovations[7].

Many other options are either in the queue or already being sold. Try to imagine what the "fax machine" or "cell phone" of the power industry will look like. As we have seen in other industries, the combination of several new technologies often results in devices few of us could have imagined only a few years ago.

WHO OFFERS THESE OPTIONS?

Accompanying the profusion of technical choices will be a more bewildering expansion of vendor choices. Even as the merger and acquisition mania gripping the industry creates new firms out of old ones, one sees a coalescence of energy services providers (ESPs) into a few distinct groups.

- Unregulated utility subsidiaries; e.g., Cinergy, Southern Electric Corp.

- Independent power providers: e.g., Sithe Energies, Gas Energy Inc.

- Mega-wholesalers: e.g., Enron, Duke Energy

- Equipment vendors: e.g., Johnson Controls, Honeywell

- Gas marketers: e.g, Eastern Energy Marketing, ERI Services Inc.

- Existing ESCos: e.g., Xenergy, EUA Cogenex

- Aggregators: e.g., New Energy Ventures, Wheeled Electric Power Co.

- Financial firms: e.g., Goldman-Sachs, Merrill Lynch

Of course, local utility distribution companies (UDCs) are also trying to stay afloat and retain their load.

THE COLLEGE OF POWER KNOWLEDGE

How does one cope with this continuously changing panorama? Fortunately, both the advancing energy industry and other innovations are providing some of the means to do so.

Getting Up to Speed. Start by learning the "lingo" and concepts. There's no need to enroll in college (and none of them teach this stuff). Most PUCs provide readable summaries of their decisions (both on paper and on their Web pages), and a variety of newsletters and free magazines are available to keep abreast of the latest changes (see appendix 1 for a list). Computer-savvy managers can "surf" informative Internet sites for even quicker access (see appendix 2 for another list).

Attending a technical conference focused on power supply issues can be very helpful to get your questions answered (see appendix 3 for a list of seminar providers). Such events are also a good way to make useful contacts. Local trade associations often sponsor panel discussions on such issues, or will hold them if interest is expressed by their members.

Speeding Up the Process. You may already belong to a trade or professional organization that is (or could be) taking action toward deregulation. Several local BOMA (Building Owner's and Manager's Association) chapters, for example, are already actively pursuing power issues. To properly represent your company's interests, membership in a customer group—or working through an energy "partner"—(i.e., a consultant or marketer) involved in rate proceedings can also be of great value. Your PUC can provide lists of past intervenors. Nationally, ELCON (Electricity Consumers Resource Council, in Washington, D.C.) represents many large industrial firms, and is a good source for user-friendly information.

But all the preparation in the world does no good unless your PUC or state legislature takes the right action on this issue. Experience has shown that only intervention in the process can make that happen. While the better marketers are already involved in this process (and those that aren't don't deserve your business), customer input is essential to ensure that the results are acceptable.

Waiting on the sidelines for "the other guy," or the PUC, to release you from your utility's grip will only prolong the present situation. Trusting your utility to do the right thing (by reducing its profit margins, selling off its assets, cutting its staff and perks) is naive: no industry has ever done so without the push of competition. Watching others bear the cost of interventions, while you reap the benefits, might give you a free ride on others' success, but experience has shown that utilities take advantage of such apathy by dragging out proceedings long enough to exhaust opponents' financial resources.

On the other hand, like many customers around the country, you can help make the right changes happen by supporting intervention into the regulatory process. When a group of energy users financially sustains such actions (either directly or through an energy partner), the contribution of each is small compared to the value of hastening competition. The payback period of such efforts is typically measured in *weeks*, not years.

Become part of the changes already underway, because those who understand the opportunities will reap the benefits of that knowledge, for both their facilities and their careers!

APPENDIX 1—free magazines focusing on the competitive utility marketplace (request subscription card)

Energy Buyer
Christine Strobel, editor
Infocast, Inc.
13715 Burbank Blvd.
Sherman Oaks CA 91401
ph: 818-902-5400
fx: 818-902-5401

MegaWatt Markets
Randy Rischard, managing editor
Pasha Publications
1616 Ft. Meyer Drive Suite 1000
Arlington VA 22209
ph: 703-816-8626
fx: 703-528-4296

PowerValue
Greg Porter, publisher
Intertec International Inc.
2472 Eastman Avenue, Bldg. 33
Ventura CA 93003
ph: 805-650-7070
fx: 805-650-7054

APPENDIX 2—Web sites addressing deregulation issues

Strategic Energy Ltd.
http://www.sel.com

Welcome to Convergence Research
http://www.converger.com/

The MCGI Home Page
http://www.mcgi.com/

Electric Restructuring in California
http://www.cpuc.ca.gov/elec.shtml

Newspage for Retail Wheeling
http://www.newspage.com/NEWSPAGE/cgi-bin/walk.cgi/
NEWSPAGE/info/d13/d4/d10/

Electric Utility Information
http://home.ptd.net/~sjrubin/electric.htm

NARUC Home Page
http://www.erols.com/naruc/

The Utility Connection
http://www.magicnet.net/~metzler/index.html

Energy Central Home Page
http://www.energycentral.com/

The National Council on Competition and the Electric Industry
http://www.erols.com/naruc/nccei.htm

Utility Deregulation Project
http://www.me3.org/projects/dereg/

The Power Marketing Association
http://www.powermarketers.com

New York State Public Service Commission
http://www.dps.state.ny.us/

misc.industry.utilities.electric Web Site
http://www.webfeats.com/preecs/miue/

The Electric Utility WWW Resource List
http://sashimi.wwa.com/~merbland/utility/utility.html

Energy OnLine
http://www.energyonline.com/

The World-Wide Web Virtual Library: Energy
http://solstice.crest.org/online/virtual-library/VLib-energy.ht ml

Energy Yellow Pages
http://www.ccnet.com/%7Enep/yellow.htm

Power Providers
http://www.powerproviders.com/

Electricity OnLine
http://www.electricity-online.com/

PEAR's Electric Intelligence: Insights on Competition
http://www.peartree.com

LEAP Letter (paid newsletter on restructuring)
http://www.spratley.com/leap

New Energy Ventures (NEV)
http://www.newenergy.com

California Energy Institute (publications page)
http://www-path.eecs.berkeley.edu/%7Eucenergy/

A useful bulletin board that discusses these issues can be accessed
by logging on to: majordomo@aesp.org and entering SUBSCRIBE
AESP-NET. You will then automatically receive e-mail covering a
variety of energy issues, deregulation being one of them.

APPENDIX 3—alphabetical list of seminar providers

AIC Conferences
50 Broad Street, 19th Fl.
New York, NY 10004
212-952-1899
fx: 212-248-7374
http://www.aic-usa.com
Yalmaz Siddiqui

American Assoc. of Utility Marketing Executives
P.O. Box 8770
Emeryville, CA 94662-8770
510-450-1815
fx: 510-655-7887
barbarap@aaume.com (Barbara Pereira)
www.aaume.com

American Business Symposiums
60 Webster Road, Suite 300
Weston, MA 02193
617-736-0800
fx: 617-736-0844

Association of Energy Engineers
4025 Pleasantdale Rd.Suite 420
Atlanta GA 30340
770-447-5083 X211
fx: 770-446-3969
www.aeecenter.org

Camber Corporation (DOE contractor)
601 13th St., NW, Suite 350 North
Washington, DC 20005
202-737-1911
fx: 202-628-8498

Center for Business Intelligence
70 Blanchard Road, Suite 4800
Burlington, MA 01803
800-767-9499
fx: 617-270-6216
registrar@cbinet.com

Chartwell, Inc.
1900 Emery Street, Suite 332
Atlanta, GA 30318
800-432-5879
fx: 404-352-8016
utilityinfo@chartwellinc.com

Clemson University
Office of Professional Development
P.O. Box 912
Clemson, SC 29633-0912
864-656-2200
fx: 864-656-0938
Amy Wright

Economics Resource Group
1Mifflin Place
Cambridge, MA 02138
617-491-4900
fx: 617-576-3514

Energy NewsData
117 Mercer Street
Seattle, WA 98119
206-285-4848
fx: 206-281-8035
newsdata@newsdata.com
www.newsdata.com/enernet

Energy Institute (Energy Seminars, Inc.)
2001 Holcombe Blvd., Suite 806
Houston, TX 77030-4214

888-353-7451
fx: 713-797-0144
nrginst@aol.com
Joshua Schwager (202-986-6746)
www.obnm.com/theenergyinstitute

Energy User News (Chilton Company)
Mike Randazzo,managing editor
201 King of Prussia Road
Radnor, PA 19089
610-964-4223
fx: 610-964-4647
mrandazz@chilton.net
www.energyusernews.com

E-Source
1033 Walnut Street
Boulder, CO 80302-5114
303-440-8500
fx: 303-440-8502
ndoty@esource.com (Nancy Doty)
www.esource.com

Infocast
13715 Burbank Blvd.
Sherman Oaks, CA 91401
818-902-5400 X22
fx: 818-902-5401
103116.625@compuserve.com (Jim Naphas)

Insight Information Inc.
55 University Ave., Suite 1700
Toronto Ontario M5J 2V6
CANADA
416-777-1242
fx: 416-777-1292

Institute for International Research
708 Third Avenue, 2nd fl
New York, NY 10017-4103
800-999-3123
fx: 212-661-6677
us002506@interramp.com (Cheryl Fallick)

International Business Communications
IBC USA Conferences Inc.
225 Turnpike Road
Southborough, MA 01772-1749
508-481-6400
fx: 508-481-7911

International Exposition Company
15 Franklin Street
Westport, CT 06880
203-221-9232
fx: 203-221-9260

International Quality and Productivity Center
150 Clove Road
P.O. Box 401
Little Falls, NJ 07424
800-882-8684, 201-256-0211
fx: 201-256-0205
lmoran@planet.net (Linda Moran)
www.iqpc.com/

King Publishing Group
627 National Press Bldg.
Washington, DC 20045
202-662-8565
fx: 202-662-9719
kingpub@access.digex.net

Pasha Publications
13111 Northwest Freeway, Suite 230
Houston, TX 77040

713-460-9200
fx: 713-460-9150
clouser@pasha.com (Gary Clouser)
www.pasha.com

Pennwell Conferences & Exhibitions
3050 Post Oak Blvd. Suite 205
Houston, TX 77056
800-883-8189
fx: 713-690-5674
umbrella@pennwell.com (Beth Baker)
www.pennwell.com

Power Marketing Association
1519 22nd St. S-200
Arlington, VA 22202
703-892-0100
fx: 703-979-4677
keltys@erols.com (Peter Dykhuis)
www.powermarketers.com

Princeton Energy Programme
136-230 Main Street
Princeton Forrestal Village
Princeton, NJ 08540-9759
609-520-9099

RER
12520 High Bluff Drive, Suite 220
San Diego, CA 92130-0081
619-481-0081
fx: 619-481-7550
www.rer.com

Strategic Research Institute
500 Fifth Avenue, 11th floor
New York, NY 10110
800-599-4950
fx: 212-302-9850

Univ. of Wisconsin Management Institute
University of Wisconsin-Madison
Grainger Hall, 975 University Ave.
Madison, WI 53706-1323
800-292-8964
fx: 608-265-3357

ZE PowerGroup Inc.
Unit #130 5920 No. 2 Road
Richmond British Columbia V7C 4R9 CANADA
604-244-1472
fx: 604-244-1675
zelramly@direct.ca (Zak El-Ramly)
www.ze.com/ze

FOOTNOTES

1.   *MegaWatt Daily*, June 27, 1997, Pasha Publications, Houston, TX.
2.   "Moving to Competitive Utility Markets: Parallels with the British Experience," by Dr. George Backus and Susan Kleeman, in March/April 1997 *PowerValue* magazine, published by Intertec International Inc, Ventura, CA,
3.   "The Competitive Effects of Transmission Capacity in a Deregulated Electricity Industry," by Severin Borenstein, James Bushnell, and Steven Stoft, published by the University of California Energy Institute, Berkeley, CA, April 1997
4.   Planergy, Inc. Web Page, http://www.planergy.com, Austin, TX, February 1997
5.   Capstone Turbine Corporation Web page, http://www.capstoneturbine.com, April 1997
6.   Product Announcement by SouthWire Corp., *Power Daily*, McGraw-Hill Publishing Co., New York, NY, April 8, 1997
7.   Entegrity Packaged Rooftop Air Conditioner product brochure, Entergy Inc., Memphis, TN November 1996

Index